Grassroots
Rising

Other Books by Ronnie Cummins

Children of the World

Genetically Engineered Food: A Self-Defense Guide for Consumers

Praise for *Grassroots Rising*

"The world is finally waking up to the ecological and climate emergency and the urgent need to realign the economy with the laws of ecology. In *Grassroots Rising*, tireless activist Ronnie Cummins outlines how we can address the multiple crises of our time by making a transition from industrial agriculture and food systems to a regenerative agriculture that recycles carbon and nitrogen in forests, grasslands, and farms; reverses climate change; creates healthy soils; and halts species extinction and the erosion of biodiversity. This is a book that should be in the hands of every activist working on food and farming, climate change, and the Green New Deal."

—VANDANA SHIVA, scientist, environmentalist, social activist;
author of *Earth Democracy*, *Soil Not Oil*, and *Stolen Harvest*

"*Grassroots Rising* is one of the most important books you will ever read. It shows the existential environmental and health disasters caused by the toxic and degenerative practices of the poison cartels, Big Agriculture, the fossil fuel industries, predator tycoons, and the money manipulators. Most importantly, though, it is a book with good news! It outlines a logical and very achievable pathway for how we can shift from degeneration to regeneration and make this a healthy, fair, prosperous, diverse, democratic, and environmentally robust world for all of us."

—ANDRÉ LEU, former president, International Federation
of Organic Agriculture Movements (IFOAM);
author of *The Myth of Safe Pesticides*

"Healthy soil, healthy plants, healthy animals, healthy food, healthy people. My friend and fellow activist Ronnie Cummins makes a strong case for how organic and regenerative food, farming, and land use can lead us to a healthier and happier world with a stable climate."

—DR. JOSEPH MERCOLA, founder, Mercola.com

"The future doesn't have to be gloomy. In this upbeat call to action, Ronnie Cummins, longtime campaigner for healthy food and land, guides us on a hopeful and pragmatic journey into the crucial upcoming decade. A Regenerative Economy is not pie-in-the-sky; it can be done. We have all the tools we need. And thanks to leaders such as Ronnie, we know what to do. This book is a must-read!"

—COURTNEY WHITE, author of *Grass, Soil, Hope* and
Two Percent Solutions for the Planet

Grassroots Rising

Rising

A Call to Action on Climate, Farming, Food, and a Green New Deal

Ronnie Cummins

Chelsea Green Publishing
White River Junction, Vermont
London, UK

Project Manager: Patricia Stone
Project Editor: Benjamin Watson
Copy Editor: Nancy Ringer
Proofreader: Deborah Heimann
Indexer: Shana Milkie
Designer: Melissa Jacobson

Printed in the United States of America.
First printing January 2020.
10 9 8 7 6 5 4 3 2 1   20 21 22 23 24

Our Commitment to Green Publishing
Chelsea Green sees publishing as a tool for cultural change and ecological stewardship. We strive to
align our book manufacturing practices with our editorial mission and to reduce the impact of our business enterprise in the environment. We print our books and catalogs on chlorine-free recycled paper,
using vegetable-based inks whenever possible. This book may cost slightly more because it was printed
on paper that contains recycled fiber, and we hope you'll agree that it's worth it. *Grassroots Rising* was
printed on paper supplied by Sheridan that is made of recycled materials and other controlled sources.

Library of Congress Cataloging-in-Publication Data
Names: Cummins, Ronnie, author.
Title: Grassroots rising : a call to action on climate, farming, food, and a green new deal /
    Ronnie Cummins.
Description: White River Junction, Vermont : Chelsea Green Publishing, 2020.
    | Includes bibliographical references and index.
Identifiers: LCCN 2019049657 (print) | LCCN 2019049658 (ebook) |
    ISBN 9781603589758 (paperback) | ISBN 9781603589765 (ebook)
Subjects: LCSH: Climate change mitigation—United States. | Environmentalism—United States.
Classification: LCC TD171.75 .C86 2020 (print) | LCC TD171.75 (ebook) |
    DDC 363.738/740973—dc23 .
LC record available at https://lccn.loc.gov/2019049657
LC ebook record available at https://lccn.loc.gov/2019049658

Chelsea Green Publishing
85 North Main Street, Suite 120
White River Junction, VT 05001
(802) 295-6300
www.chelseagreen.com

CONTENTS

Introduction

regeneration. *1: an act or the process of regenerating: the state of being regenerated. 2: spiritual renewal or revival. 3: renewal or restoration of a body, bodily part, or biological system (such as a forest) after injury or as a normal process.*

MERRIAM-WEBSTER'S COLLEGIATE DICTIONARY, 11th ed.

This is a book about how we—the United States and a global grassroots movement—can rise up together and overcome the most serious threat that humans have ever confronted: global warming and severe climate change. According to scientific consensus, unless we can begin to rapidly turn things around, within one to two decades, our current climate crisis will likely morph into catastrophic climate change, unraveling the life-support systems of our planet.

The driving force that informs and inspires our new grassroots revolution is Regeneration, a rapidly spreading, carbon-sequestering, ecologically restorative, technologically innovative, forward thinking worldview that takes us well beyond the now unfortunately outdated twentieth-century notions of sustainability.[1] Regeneration calls for a transition from degenerative, climate-disrupting fossil fuels to renewable energy and from industrial food, farming, and land use to regenerative practices. In this way, through the miracle of plant photosynethsis, we can draw down billions of tons of excess carbon from the atmosphere into our soils, forests, and plants over the next few decades and thereby avert climate catastrophe. By mobilizing the grassroots power of a united body politic for survival and revival, we can head off climate chaos and build a new nation along the lines of a Green New Deal. At the same time, as our Regeneration revolution spreads across borders, we can build a new global commonwealth of peace and justice.

A properly organized and executed Regeneration revolution, led by global youth and a revitalized US and global grassroots not only has the awesome capacity to draw down massive amounts of excess atmospheric carbon dioxide (CO_2) and reverse global warming but, at the same time, has the power to clean up pollution, restore water quality, increase biodiversity, and rejuvenate soils, forests, pasturelands, croplands, wetlands, and watersheds. Moreover, this revolution in our relationship to Mother Earth and one another, scaled up nationally and internationally, has the potential to revitalize public health, both mental and physical, by providing a bountiful harvest of healthy, organic foods for everyone, while transforming our currently degenerative urban and rural landscapes into regenerative environments and bringing us all together in a common mission. Coupled with a green and equitable energy economy, regenerative agricultural and land use policies and practices hold out the promise of a better life and standard of living for all of the world's 7.5 billion people, rural and urban, including the most impoverished and exploited communities.

You're holding in your hands an international declaration of emergency, a "bad news" chronicle of the near-terminal damage that fossil fuels, out-of-control commerce and greed, corrupt politics, and degenerative food, agricultural, and land use practices have inflicted on Mother Earth and society. But you're also holding in your hands a "good news" instructional book for Regeneration, a practical, shovel-ready plan of action for the United States and the world to transition to climate stability, peace, justice, health, prosperity, cooperation, and participatory democracy.

My motivation for delivering the "bad news" portion of this book comes from my rage and alarm at the steady deterioration of the United States and the world, now teetering on the brink of catastrophe. But what inspires me more nowadays is the emerging "good news," the heretofore lesser-known but positive and redemptive message that we, the global grassroots, can regenerate Earth and put an end to "business as usual": the climate chaos, poverty, forced migration, deteriorating human health, environmental destruction, and endless wars that are no longer tolerable. Properly launched and scaled up, from Main Street to the Middle East and beyond, the Regeneration revolution we are proposing has the potential to cure our depression, revitalize our moribund politics, and restore our sense of global solidarity.

I, like so many Americans, unfortunately, am somewhat of an expert on *degeneration*. I grew up in southeast Texas on the edge of the Gulf of Mexico, in a hot and humid coastal wetland, rich with birds and wildlife, mosquitoes, snakes, and alligators, teeming with fish, crabs, and shrimp, where few Europeans, indigenous people, and African Americans had settled before the "Spindletop gusher," the largest oil deposit in the United States, was discovered in 1901.

By the time I was born, just after World War II, the Golden Triangle, as the oil industry and local chamber of commerce liked to call the area that included Port Arthur, Orange, and Beaumont, Texas, had become the largest complex of petrochemical plants in the world, employing thousands of workers, many of whom belonged to the Oil, Chemical, and Atomic Workers (OCAW) trade union. My dad was one of them. After returning from World War II, he worked at Gulf Oil for almost fifty years, punching the clock at the same refinery where his father had worked for forty years.

My mom's parents, on the other hand, were decidedly more rural folk. After working city jobs in the Golden Triangle for a number of years, they moved back to the land in the mid-1950s, operating a diverse and old-fashioned family farm—an organic farm in reality, although they didn't use that word to describe the traditional practices that they had learned from their Cajun parents and grandparents in southwestern Louisiana. I spent idyllic weekends and summers on my grandparents' farm in East Texas, taking care of the farm animals (I can still remember the names of my favorite milk cows, January and Bossie, and our pet pig, Cliff), gathering eggs from the chicken house, helping out in the vegetable gardens, roaming the piney woods with my border collie, and diving off the railroad trestle with my brother, sisters, and cousins into the sandy spring-fed creek nearby, where the water flowed clear and cold.

But the post–World War II America of my childhood now seems like a fairy tale, an old TV episode of *Leave It to Beaver*, *Bonanza*, or *Ozzie and Harriet*. As I grew older, the nearby Gulf of Mexico and our bayous and rivers became more and more polluted. At the jetty where my brother and I had once caught fish, crabs by the bushel basket, and shrimp, the water became fouled with oil, green algae, and a nasty collection of floating plastic and garbage. After a while my brother and I stopped fishing and crabbing. More and more sticky black tar began washing up on the sands of McFaddin Beach, our favorite swimming spot. To get the tar off the soles of our feet, we'd have to dip a rag in gasoline or paint thinner and use it to rub our skin until it turned bright red.

At dusk and throughout the night, the sky throughout the Golden Triangle glowed fluorescent orange, eerily tinted by the methane and toxic gases being flared off 24/7 by the oil refineries. The air in my neighborhood, in the schoolyard, at the Little League baseball diamond, and even downtown smelled worse and worse. In high school I finally stopped going outside for lunch with my friends because the air was so foul. The chemical plant down the road was later classified as one of the most toxic industrial facilities in the nation.

Having spent most of their lives surrounded by oil refineries and chemical plants, most of my high school classmates died prematurely, many from cancers associated with environmental toxins, polluted drinking water, and low-grade, highly processed poison food. Today, the memory of those who died too soon fuels my passion as an organic food and environmental campaigner against the Poison Cartel—Bayer/Monsanto, Dow, DuPont, Syngenta/ChemChina, ExxonMobil, BP, Koch Industries, Halliburton, and all the rest—a cartel that, as you probably know, has polluted not only my hometown but the entire world.

The degeneration zone of my youth, which the media has aptly dubbed "Cancer Alley" (denoting the corridor between Houston, Texas, the Golden Triangle, and New Orleans), has been ravaged not just by industrial pollution but also by political corruption, economic disparities, and racial strife. That combination comes together with devastating results in the face of one of the biggest, most threatening forces along the Gulf Coast: hurricanes. Several years ago I watched the TV news coverage of Hurricane Harvey. Among the clips were shots of my hometown, under water, including astounding footage of several elderly nursing home residents in wheelchairs, filthy floodwater rising up to their laps. Hapless Houston and Gulf Coast residents, typically working-class whites, Latinos, and African Americans, complained to the news media during and after Hurricane Harvey of refinery explosions, toxic chemical releases, and poisonous floodwater, but oil company spokespersons and elected public officials said more or less the same thing they've been saying since DDT's developer was awarded the Nobel Prize in 1948, since Hurricane Katrina swamped New Orleans in 2005, and since the BP Deepwater oil spill in 2010 ravaged the Gulf: "No need to worry."

In 2016 I helped raise money and recruit volunteers for Standing Rock, the Native-led protest and encampment in North Dakota mobilizing to stop the infamous Keystone XL pipeline. Part of my zeal for stopping this pipeline, which was designed to ship billions of gallons of dirty tar sands oil from Alberta, Canada, to the United States for refining, came from its ultimate destination: Port Arthur, Texas. Port Arthur specializes in refining dirty oil, no doubt because industry can get away with literal murder in the National Sacrifice Zone between southeast Texas and coastal Louisiana.

Growing up in Texas, in a company town owned lock, stock, and barrel by the oil giants, I was always interested in politics. Because my father was in the OCAW trade union at the oil refinery, an integrated union with whites, blacks, and Latinos as members, I learned early on that racism was not only morally wrong but an impediment to democracy and working-class unity. Unlike many Southern whites at the time, my parents and grandparents were respectful of other races and nationalities, politely referring to blacks as "colored people" or Negroes.

My hometown was completely segregated at the time. There were no blacks in the schools, churches, civic clubs, or any of the white neighborhoods. Restrooms and drinking fountains in public places were labeled as "whites" or "coloreds only." Blacks were forced to live in the ghetto of Port Arthur, across the railroad tracks, many in ramshackle houses straight out of the antebellum South of the nineteenth century. They were basically forbidden to come into the white part of town after dark. Those caught on the wrong side of the tracks after sunset ran the risk of being accosted, not only by the all-white police force, but by racist gangs as well. Because of the intense segregation in the Golden Triangle, I never really met any black people my own age until I went off to college and got involved in the anti-war and civil rights movements.

I was (and still am) a voracious reader, especially of history books and novels with political themes. In high school, I was intrigued by the radical populist movement of the late nineteenth and early twentieth century; as it turns out, the Farmers' Alliance, an agrarian populist organization, had been especially strong in Texas, Oklahoma, and Louisiana. I was fortunate then to have a young civics teacher, fresh out of college, who taught us all about the Abolitionist movement and the Civil War, the robber barons of the nineteenth century, the populist movements of the late 1800s, the Depression and the New Deal, and the civil rights and anti-war movements that were just then starting to gain momentum. His influence crystallized some of the ethical and political views that my grandparents and father had tried to instill in me.

When I was nine years old, my Cajun grandparents took me on a sightseeing trip to the Louisiana State Capitol in Baton Rouge. In a scene that I will never forget, my grandfather, or Papa, as I always called him, led me over to the marble wall behind the Speaker's podium in the room where the state legislature meets. He put my fingers in several of the numerous pockmarks in the wall.

Papa said, "Son, these are bullet holes from a machine gun. This is where the oil company men, Rockefeller's gunmen, took down Governor Huey Long on September 10, 1935."

I said, somewhat stunned, "Papa, why would they kill the governor?"

He replied, "Because Huey was organizing the poor people, black and white, small farmers and sharecroppers, to take back what was rightfully ours."

My life experience has taught me that money rules and power corrupts, and that putting profits before people and environmental health is not only wrong but deadly. But fifty years of activism—from the 1960s radical student movement through the modern ecology movement that emerged in 1970, the anti–nuclear power movement of the late 1970s, the Central America solidarity movements of the 1980s, and the organic, climate change, and Occupy movements of the past twenty-five years—has also taught me that organized grassroots power can make a big difference, whether we're talking about public consciousness, marketplace pressure, or politics and public policy. The current situation we are facing in the United States and throughout the world is indeed dire, on all fronts, and time is running out. But as I and my fellow activists in the new Regeneration movement believe, there is still time to turn things around, before it's too late.

1

Rules for Regenerators

Hope is something that you create, with your actions. Hope is something you have to manifest into the world, and once one person has hope, it can be contagious.

ALEXANDRIA OCASIO-CORTEZ,
in "When Alexandria Ocasio-Cortez Met Greta Thunberg"

Over the past five decades, as a food, natural health, and environmental campaigner, anti-war organizer, human rights activist, and journalist, I have had the unique, inspiring, and, at times, depressing opportunity to work and travel across much of the world, in both rural and urban areas and in developing and "overdeveloped" countries alike. From the anti-war, civil rights, and anti-nuclear movements of the 1960s and '70s to the war zones of Central America in the 1980s, the street protests and public interest campaigns of Washington, D.C., in the 1990s, and my current US and international work on food, farming, health, and climate, I have seen some of the best and worst examples of human behavior and collective action, including a rather full spectrum of what I call "activist malpractice."

After delivering thousands of talks and presentations, organizing rallies, boycotts, and protests, doing media interviews, writing articles, lobbying politicians, raising millions of dollars every year in donations from small and larger donors, and helping build local, national, and international coalitions, perhaps the most important thing I have learned is that people respond best to a positive, solutions-oriented message. Gloom-and-doom messaging—whether we're talking about climate change, poverty, health, war, or political corruption—that offers no plausible solution does not generally inspire people to get involved or take action. On the other hand, given the seriousness of our current situation, it's important not to downplay the unprecedented life-or-death threats that we face, nor the formidable political, economic, and cultural obstacles that block our way forward. Our new Regeneration movement needs to be able to communicate the absolute seriousness of our climate emergency and societal degeneration, but at the same time, it must deliver the good news that we have the tools, growing grassroots support, and a practical plan, embodied in the Green New Deal in the United States (see chapter 5), to turn things around.

Yes, it's extremely important to highlight and criticize, with passion, facts, and concrete examples, the bad actors, practices, and policies in our contemporary world. We must continue to expose and analyze the machinations of the billionaire Degenerators, the global ruling elite. We must expose and fight against the destructive behavior and criminality of out-of-control corporations, such as Bayer/Monsanto, Walmart, McDonald's, Cargill, Amazon, Facebook, Google, Microsoft, Fox News, and ExxonMobil, as well as the antics of indentured and corrupt politicians. We must steadily educate and mobilize a mass base of support—consumers, students, farmers, workers, politicians, activists, and whoever else will listen—and defend ourselves from the ongoing attacks on our biological life-support systems. We must expose dangerous practices such as factory farming, GMOs, chemical-intensive agriculture, geo-engineering, fracking, and all forms of environmental destruction.[1] But whether we're talking about fossil fuels, climate change, factory farms, deteriorating public health, corporate crime, poverty, war, or corrupt government, the most effective critique that resonates with people, with hearts and minds, is a narrative that offers positive solutions.

Once we are able to connect the dots and understand the issues in a personal way, and once we are approached by organizers whom we know and trust—organizers who can offer up a plausible plan—most of us are willing to join up or take action. This is especially true if we understand that we are dealing with a genuine state of emergency and we have heard and believe, or better yet, we have seen with our own eyes, that there are practical solutions to the problem.

As the roving international director of the Organic Consumers Association and a steering committee member of Regeneration International, I am privileged to travel around the world and work on a daily basis with staff, affiliates, and everyday grassroots activists in over thirty countries in

North and South America, Europe, Africa, Australia, and Asia: farmers, consumer organizers, political activists, scientists, natural health advocates, journalists, and more. My various "home offices" bring me into close contact with a number of radically different environments and communities: the rural far north woods of Finland, Minnesota; the shores of Lake Superior, just south of the Canadian border; the progressive, decidedly cosmopolitan city of Minneapolis, Minnesota; the megalopolis and political and environmental hot spot of Mexico City; and the high desert municipality of San Miguel de Allende, Mexico, in the state of Guanajuato, where I live and work for several months each year on Regeneration International's organic farm school and conference center outside of town and help manage an organic grocery store, restaurant, and educational center.

My window on the world has developed through my work as a "content scout" for our organizations' websites and social media sites. My daily task for the past twenty-three years has been to read, review, aggregate, and post a never-ending stream of articles, research papers, interviews, and videos, with a focus on food, farming, land use, natural health, climate, and politics, paying attention to both the bad news and, more importantly, the good news, in terms of trends and practices.

The global phenomenon that encourages me most nowadays is that a growing corps of youth, women, indigenous communities, consumers, farmers, and traditionally oppressed people are waking up, connecting the dots, and starting to get organized, even if this peaceful global grassroots rising is not yet adequately covered by the mass media nor understood by the general public. The ripening of objective and subjective conditions becomes especially clear to me when I give a talk or write an article connecting the dots between regenerative food, farming, land use, climate change, and other pressing issues. A critical mass of people, and especially young people, not only are open to the good-news message that renewable energy and regenerative food, farming, and land use can positively resolve the climate crisis and other burning issues but are ready to join up and become part of this new revolutionary movement.

Based upon my own daily experiences and those of the people whom I rely upon in my life and my work, I have come to believe that the main obstacle we have to overcome, in the United States and worldwide, is that many, if not most, people are locked into disempowering work, school, and social situations. Most people are stranded in familial and social circles among people who are pessimistic about significant social change. Consequently, people suffer from a pervasive sense of hopelessness, depression, loneliness, and alienation. They spend ten hours a day or more staring at computer screens or smartphones, instead of positively engaging with real people and participating in community-building or social change movements.

According to the health insurance company Blue Cross Blue Shield, among its forty-one million insured members in the United States, diagnoses of psychological depression have increased by 33 percent in the past five years alone, with an estimated nine million Americans now suffering from major depression. Incidence of major depression is increasing most rapidly among youth and adolescents, especially young women.[2] The grind of daily survival, boring and meaningless jobs and social roles, irrelevant educational institutions, and a pervasive lack of awareness and solidarity, especially international solidarity, add up to an overall sense of powerlessness: you can't fight City Hall, the federal government, and big corporations. Many, if not most, people have doubts that Big Oil, Big Corporations, and corrupt politicians can be brought to heel in time to turn our global crisis around. In most cases no one has ever seriously tried to explain to them or show them that another world *is* possible and that there *are* systemic solutions to all of the problems that we face, including the climate crisis.

A growing number of people are well informed and increasingly alarmed about climate change, but most people don't understand that we can actually *reverse* global warming, as opposed to just slowing it down. They're familiar with solar, wind, and other forms of renewable energy and the necessity of rebuilding our national and international infrastructure to drastically reduce fossil fuel emissions, but they're not fully convinced that we can achieve net zero fossil fuel emissions by 2030 or even by 2050, as scientists warn us that we must. Most people have never heard about, much less seen for themselves, the amazing power of regenerative food, farming, and land use to sequester atmospheric carbon and store it in our soils, forests, and plants. Most people (even climate activists) have typically never heard about the incredible potential of healthy soils, plants, and trees to naturally sequester and store in our soils enough excess carbon from the atmosphere—adding up to billions of tons every year—not only to mitigate climate change but to reverse global warming. Even fewer understand that regenerative organic farming and ranching can also increase the nutritional density of everyday foods enough to rejuvenate public health, lift hundreds of millions of small farmers and rural villagers out of poverty, dramatically reduce forced migration, solve the water crisis, preserve biodiversity, and alleviate the major causes of war and terrorism.

As Native American Chief Seattle reportedly said, "Without a vision, the people will perish." It's not that people are satisfied with their current situation and don't want things to change. The problem is that *most people don't really believe things can change.* Therefore, our central task as Regenerators is to shift the global conversation on food, farming, politics, health, and climate from one of hopelessness to one of hope, and to empower the grassroots to rise up and take action, both individually and collectively. That's what this book is all about.

If we are going to turn things around and essentially build a new system, then we have to convince people, as the slogan of the World Social Forum puts it, that "another world is possible." And we need to prove this, not just in theory, but in everyday action, drawing upon successful real-world, shovel-ready, tried and proven practices, whether we're talking about renewable energy and energy conservation or regenerative food, farming, and landscape management. We need to publicize and showcase the renewable and regenerative projects, practices, and policies that already exist in every region and ecosystem in the world, even if these game-changing energy practices, buildings, farms, ranches, food hubs, educational projects, and natural health practices are not, for the moment, widely known.

People are depressed, angry, and alarmed about the state of the world, but they're not sure what to do about it. Therefore, rather than argue about abstract ideology or debate about which of our single issues are most important, as many people have done for the past fifty years, our primary task today is to articulate and spread a positive vision of Regeneration, starting with what we eat, how we grow our food, and how we treat the land, small farmers, and farm animals. The problems we face are enormous and omnipresent. And yet the solution to these problems lies right under our feet, at the tips of our forks and knives, and in the voting booth at our local precinct. Regenerative food, farming, and land use can provide a new outlook on life, a therapeutic vision and daily practice that demonstrates that we the people, the global grassroots, can begin to turn away from disaster, solve our most pressing crises, and meet our most important needs.

For this reason, I'm not going to spend a lot of time going over all the negative and depressing developments on the political, economic, and social scene that you already know about. I will focus instead on the positive, regenerative solutions that are already emerging and concentrate on how we can scale up these practices and build a global movement and green economy, across borders and continents, that will enable us not only to survive, but to thrive.

In the spirit of positive messaging, the following are six basic rules for organizing a grassroots Regeneration revolution.

Rule 1: Search Out and Emphasize the Positive

Regenerators, given our current dire situation, need to operate on the old adage that "the darkest hour is right before the dawn." Instead of dwelling on the negative, we need to search out and highlight positive trends and practices. On the contemporary scene, there are numerous signs of change and powerful countervailing trends to the degenerative status quo, not only in the United States but across the world. Rather than dwelling on gloom and doom, we need to focus on positive, potentially world-changing trends. Here are a few recent positive developments on the climate and political fronts that we can all draw upon for inspiration:

☛ Practically every government in the world, except for the Trump administration in Washington, D.C., has signed on to the Paris Climate Agreement to move to net zero fossil fuel emissions by 2050, with many nations promising to move even faster. Three dozen nations have also signed on to the lesser-known but critically important "4 per 1000" Initiative, a bold international policy initiative and agreement, complementing the global transition to 100 percent renewable energy, to draw down enough excess atmospheric carbon through regenerative food, farming, and land use practices to not only mitigate but actually reverse global warming.[3]

☛ Renewable energy already has begun to replace fossil fuels, not only in more democratic and egalitarian regions like the Nordic countries or in industrial powerhouses like Germany and the United States, but in China and India as well. It is now cheaper in many parts of the world to invest in wind and solar power than to build new coal plants. Soon it will be more profitable to install solar and wind power than to keep existing fossil fuel plants running. And it now appears that electric cars, trucks, buses, and trains have the potential to replace most gas- and diesel-powered vehicles within the next three decades.[4]

Global investments in renewable energy are currently estimated to be $300 billion a year. If we can push governments and the private sector to increase these investments to several trillion dollars per year—less than half of what the world's governments now spend on propping up fossil fuels—we should be able to make a just transition to near 100 percent renewable energy by 2030 to 2050.[5] Investors and public institutions, under the pressure of climate activists, are starting to divest trillions of dollars from the fossil fuel industry, while increasing their investments in renewable energy and conservation.[6] If food, farming, and peace activists can expand the divestment movement into pressuring investors, institutions, and everyday people to divest significant resources, not only from the fossil fuel industry but from industrial agriculture and the military-industrial complex as well, and build up political power in the United States and other nations, we can achieve our goals. With a new regenerative investment and funding network, we can free up trillions of dollars currently propping up fossil fuels, industrial agriculture, military spending, and destructive land use and reinvest this money in a complete overhaul of our food, farming, energy, transportation, manufacturing, and land use systems, providing green jobs and economic livelihoods for hundreds of millions of rural and urban workers worldwide.

☛ A critical mass of the global grassroots is starting to wake up and resist—North, South, East, and West—by organizing politically, slowly but surely developing practical and equitable solutions to our most pressing problems: climate change, poverty, injustice, environmental pollution, war, deteriorating public health, forced migration, unemployment, discrimination, and political corruption. In the United States, progressive and radical forces, led by youth, women, and minorities, appear poised, over the next decade, to sweep the majority of corrupt politicians from office, not only in the nation's forty thousand cities, towns, and counties but at the federal level as well. Similar trends are emerging in dozens of other countries, even in repressive dictatorships such as China, Russia, and Iran and in violent narco-states such as Mexico. The bottom line is that people all over the world are fed up with corrupt politicians, greedy businesses and corporations, organized crime, and a polluted environment. As the millennials are starting to understand, there is no future for their generation, nor for any of us, without fundamental change and Regeneration.

A growing global majority understand that if we wish to survive, our nations, especially the United States, simply cannot continue to maintain their respective global empires or spheres of interest or to subsidize fossil fuels, the military-industrial complex, Big Pharma, and degenerative food, farming, and land management practices. Nor can multinational corporate elites continue their "profit at any cost" practices, reinforced by hyperconsumerism on the part of consumers.

The number one national security threat for the United States and other nations today is not competition or conflict with Russia, China, India, or Iran but rather the growing danger of catastrophic climate change and the unraveling of the biological carrying capacity of Earth to support human civilization. Likewise, the biggest threat facing our counterparts in Russia, China, India, and Iran is not the United States, NATO, and the CIA but global warming and the destruction of our common home.

Instead of rallying behind politicians who demonize other nations or religions, deny that climate change or other serious global issues exist, pander to giant corporations, or harp on divisive single issues, we must collectively acknowledge that we, the global grassroots, are all one in our need for a livable environment, stable climate, equitable economy, meaningful employment, peace, and democracy. Focusing on deploying, financing, and scaling up renewable energy and regenerative food, farming, and land use systems, many of which are already in place, or partially in place, we have the power to reignite hope, activism, and participatory democracy at the global grassroots and thereby regain control over our destiny.

What we need in order to save ourselves from disaster is obviously not a classic revolutionary insurgency; we won't be preparing for a frontal assault on the Kremlin or the White House. (In the twenty-first century this type of revolution is impossible, given that we the people are now totally under surveillance, policed, and outgunned.) What we need instead is a nonviolent but determined "March of Regeneration" that moves through all of our communities and institutions, a peaceful but forceful global grassroots uprising from below in all of the 195 countries of the world.

We need a nonviolent Regeneration revolution on a global scale of public education, grassroots lobbying, marketplace pressure, and scaling up of best practices, systems, and policies, especially in the energy, food, farming, and land use sectors.

Given the overwhelming power of the establishment and its forces of intimidation and repression, we will need to carefully develop and deploy our new Regeneration paradigm or ideology like a form of jujitsu or martial art, with the power to peacefully divide, co-opt, and conquer our political and corporate adversaries, while at the same time regenerating the Earth and uniting, across issues and borders, a critical mass of the global grassroots. This is a Regeneration revolution and a new global solidarity that can move us away from impending catastrophe and give birth, as Pope Francis puts it, to "an awareness of our common origin, of our mutual belonging, and of a future to be shared with everyone."[7]

Rule 2: Link Up with People's Primary Concerns and Connect the Dots

In addition to highlighting positive solutions, it is important to keep in mind that different people have different situations, perspectives, passions, and priorities. One size or one approach does not fit all. Therefore, we need to integrate our green justice and Regeneration messages with the specific issues and concerns that are most important to grassroots constituencies and then lay out, in everyday language, a strategy that makes people understand that we can actually solve the problems they care about the most, while solving a host of other pressing problems at the same time. Only by starting from where people are at, and then connecting the dots, can we capture the attention and imagination of a critical mass of the global grassroots and get them to start thinking about how they can participate in our new movement and new economy.

Objective and subjective conditions for change are different in every one of the 195 countries in the world, and to some extent they differ as well in the subregions and local communities of these countries. Everyday people everywhere, including the most impoverished and vulnerable communities, have their burning issues as well as their secondary issues, accompanied by an inherent desire to alleviate and, if possible, solve the problems that are pressing down on them, in many cases threatening their very survival. In the activist community, major focus areas, in most cases reflecting the concerns of everyday people, include climate change, environmental pollution, health, social justice, jobs and economic justice, peace, and democracy. Unfortunately, campaigners, on both the local and national/international levels, often work in isolation from other sectors, each in their own separate silo. This perpetuates tunnel vision in the body politic, parochial or sectarian attitudes, political polarization, and an overall weakness in global civil society.

Because conditions are often so different in different communities and nations, Regenerators need to think, plan, and act strategically and holistically. If a particular community's primary concern is poverty, unemployment, racial discrimination, health issues, or substandard public schools, school lunch programs, and school curriculums—issues that are very important in frontline or impoverished urban areas such as Detroit, Michigan, or Washington, D.C., for example—then we must strive to integrate our climate and Regeneration activism with people's everyday concerns. This is why we must rally behind comprehensive, indeed system-changing political and economic programs such as the multi-issue, multi-constituency-oriented Green New Deal and not be afraid to be called "radical." To gain majority support, we must offer up a systematic solution to multiple crises, combining economic justice and jobs creation with rebuilding urban neighborhoods and rural communities, upgrading housing, providing needed social services, and moving to renewable energy. In the context of our current climate emergency and everyday economic crisis, a Regeneration organizer must be able to talk about how regenerative food, farming, and urban/rural ecosystem restoration not only can sequester carbon and help restabilize the climate but can address pressing community health, nutrition, and economic issues at the same time.

If people are passionate about addressing the air and water pollution caused by a pipeline, factory, mine, or industrial factory farm, then Regenerators must get involved, not only by trying to stop or close down the particular pipeline, factory farm, or industry, but also by projecting positive solutions, such as renewable energy, and scaling up the organic and regenerative food, farming, and land use alternatives that already exist.

Health concerns, particularly in regard to children, are a burning issue for hundreds of millions of people. One of our primary arguments as advocates for regenerative food and farming should be to talk about not just the beneficial climate and ecological impacts of regenerative agriculture but also the tremendously beneficial impacts of organic and regenerative crop production and grazing on improving food quality and nutritional density. Healthy soils and landscapes, managed in a regenerative manner by farmers, ranchers, and gardeners, give rise to healthy vegetables, fruits, grains, and animals, which in turn engender healthy food and healthy people. People need to stop eating factory-farmed meat, dairy, and poultry, not just because it's cruel to animals and bad for the climate and the environment, but also because these products are bad for your health and the health of your children. One hundred percent grass-fed meat and dairy, for example, and regeneratively produced meat, dairy, and eggs are filled with healthy omega-3 fats, vitamins, trace minerals, and other nutrients, whereas factory-farmed meats, dairy, and eggs are filled with bad fats and animal drug and pesticide residues and have lower levels of essential nutrients.[8]

———

We need to get people to understand that most of their (and our) problems and stresses are symptoms of a degenerate system that needs to be replaced, and that *can* be replaced, starting with small changes in our everyday lives and local communities that fall into line with the strategies of a larger campaign for Regeneration, such as the Green New Deal (see chapter 5). This strategic "connect the dots" approach is key to consciousness raising, coalition building, marketplace pressure, and grassroots mobilization and fundraising. This is the only way that our new Regeneration movement will be able to bring about "the development of new convictions, attitudes, and forms of life," as Pope Francis put it, that are necessary for our survival.[9]

Rule 3: Stop Organizing around Limited Single Issues

As Regenerators, we must start by meeting people where they are—that is, acknowledging and engaging with them around the issues that they most care about—and then we must help people understand the connection between those issues and the issues that most concern other people or communities who are potentially their allies. Single-issue thinking is a major form of activist malpractice that routinely gives rise to divided movements and fractured constituencies. To bring about true Regeneration, or even to pass sweeping new regenerative legislation, such as a Green New Deal, we must be not divided and fractured but united, inclusive, and holistic in our understanding of the global crisis we face.

Single-issue thinking might give rise to sentiments like "My issue is more important than your issue," "My constituency or community is more oppressed or more important than your constituency or community," or even "My solution is the only solution." In constrast, our new global Regeneration movement must be built upon the principle that *all* grassroots issues and *all* constituencies are important. Why? It's not just that strength comes from numbers. We have to help each other recognize that the burning issues bearing down on the global body politic—climate change, poverty, unemployment, declining health, political corruption, out-of-control corporations, war, and more—are the interrelated symptoms of the diseased system of Degeneration.

To solve the crisis of the system that threatens the well-being and existence of all of us, rich or poor, our movement will have to become decidedly internationalist, multi-issue, and ecumenical. Especially in the United States, the most powerful and influential nation on Earth, we can no longer afford to self-righteously wall ourselves off into our separate silos as environmental, food, health, justice, political, gender rights, animal rights, or climate activists. We cannot ignore or marginalize other issues, constituencies, and movements, whether domestic or international. We must connect the dots and all become engaged, not only in the marketplace (voting with our consumer dollars for regenerative products) and in the public square (voicing our opinions and educating our fellow citizens), but also in the political process, on the local, state, and federal levels. This holistic approach does not mean that we must give up on our primary passions and everyday concerns for survival; rather, we must weave these particular issues and our natural constituencies into a broader national and international movement.

Even though most US nonprofit public interest organizations are partly restricted by law from engaging in overt electoral/political activities (i.e., endorsing candidates, raising money for them, or spending more than a certain percentage of an organization's resources on lobbying for specific policy change or legislation), there's nothing to prevent nonprofits and public interest activists from forming separate entities or joining forces with political organizations fighting for change that are open to supporting our issues. Regenerators must strive to bring issues-oriented movements into synergy with the movements for political change, and vice versa. In the United States and elsewhere, we must draw clear connections between the most strategic issues of Regeneration and more specific constituency issues, disrupt and overturn business-as-usual policies and mind-sets, and carry out a ballot box revolution at the local, state, and national level, or we will perish.

Rule 4: Stop Pretending That Partial Solutions or Reforms Will Bring About System Change

The rise of authoritarian and fascist regimes and the weakening of a common sense of purpose, cooperation, and solidarity have brought us to a dangerous precipice. Will the United States and global grassroots wake up in time, break down the false walls between all of our burning issues, and unite across borders in a common global campaign for survival and Regeneration?

Activists fall into the trap of malpractice when they project partial solutions or tactics as if they are systemic solutions. One of the most alarming is the notion that 100 percent renewable energy will, in and of itself, solve the climate crisis. That's both misleadingly hopeful and dangerously flawed. Renewable energy will not get us to net zero emissions by 2030 or even 2050 unless it is accompanied by a *massive* drawdown (of 250-plus billion tons) of excess carbon from the atmosphere through regenerative food, farming, land use, and commerce. Both of these things—renewable energy and drawdown—need to be carried out simultaneously over the next twenty-five years.

Similarly naive, narrow-minded thinking might lead us to believe that campaign finance reform or electing a certain candidate to office will solve the national and international crisis of elite domination and political corruption, or that, in general, change in one community or country can solve what are essentially national and global problems. Unless we can lift our heads, connect the dots, and fight for unifying systemic changes, any changes that we do make won't be effective.

Climate activists have done an outstanding job in undermining the fossil fuel industry, but at the moment, the movement's leaders have not really incorporated drawdown and regenerative food, farming, and land use into their communications and campaign strategy, much less pointed out that regenerating soils, landscapes, and diets will also solve a host of other critical problems. They have not, for the most part, made the convincing case that we must communicate to the global grassroots: that we can actually reverse, not just slow down, global warming while solving related problems of rural poverty, forced migration, environmental destruction, deteriorating public health, and war. As a result, the present climate movement is somewhat isolated from other movements, such as the food, farming, natural health, labor, peace, and justice movements (to name just a few), all of which could and should be its natural allies.

At the same time, these other movements have themselves been walled off in their own separate silos, framing their own narrow field of interest as the most important for changing the system. They have failed to incorporate some of the keystones of the climate action movement—addressing the climate crisis, moving to 100 percent renewable energy, and fighting for the political change we need to make both of these things happen—into their educational message, even though it's clear that these issues are related. The continued use of fossil fuels (including their massive use in chemical-intensive agriculture, food processing, and transportation) and the domination of our political system by fossil fuel special interests will increasingly make it impossible, under ever more severe climate change conditions, to preserve our essential biological support systems, restore public health, rebuild rural communities, and grow enough food to feed the planet. Every movement—health, labor, peace, justice, environment, food—must put the Climate Emergency at the top of their agenda.

Similar activist malpractice and single-issue thinking have come from often well-meaning but misinformed animal rights and environmental advocates who routinely demonize livestock, ranchers, and meat eaters as the primary culprits in climate change, environmental destruction, and deteriorating public health. While it's true that factory farms and the GMO grain monocultures that support them, along with the consumption (indeed, overconsumption) of factory-farmed meat, dairy, and poultry, are a primary driver of all these problems, that's not the full story.[10] A comprehensive solution to these problems requires not just an end to factory farms and a global rejection of factory-farmed foods by consumers but also regenerative grazing (neither over- nor undergrazing) and animal husbandry practices. Properly (regeneratively) raised livestock have the awesome power, through the sustenance they provide in the great chain of biodiversity, not only to provide food and nutrition for humans, but also to naturally fertilize the soil, prune the grasses, increase plant photosynthesis, eliminate invasive plant species, increase rainfall capture and infiltration in the soil, increase aboveground biomass and underground biological life, prevent erosion, and sequester billions of tons of atmospheric carbon every year.[11]

Our planet holds eight billion acres of grasslands, pastures, and rangeland, most of which is not suitable for growing crops, most of which needs healthy populations of trees and perennials so as to optimize carbon sequestration, and most of which will benefit from regenerative grazing. Other benefits accrue as well. In addition to fostering healthy ecologies, practicing regenerative grazing on a global scale will require that we mobilize several hundred million small-scale ranchers, farmers, and herders, all working on a modest scale, and all earning a livelihood. It will require support from not only green-minded governments but also an army of conscious consumers willing to pay a fair price for meat, dairy, and eggs that are produced in a climate-friendly, humane manner and that, in return, maximize human and environmental health.

Returning to our main point, though, we must avoid thinking about any single issue as a complete solution to the crises facing our world. While regenerating the world's eight billion acres of grasslands, rangelands, and pastures is essential to carbon sequestration, that in itself is not sufficient to reverse climate change, given the supersaturation of the atmosphere with CO_2 and other greenhouse gases. Other land-management strategies include regenerating our four billion acres of croplands and restoring, reforesting, and regenerating our ten-plus billion acres of forests, wetlands, and peat bogs.

In a similar vein, natural health advocates and nutritionists, while calling for an end to factory farms and promoting a diet that is predominately plant-based, need to point out that grass-fed, pastured, and organically raised meat, dairy, and poultry and wild-caught fish such as salmon are very healthy foods, when consumed in moderation. Strict veganism, on the other hand, can lead to nutritional deficiencies and chronic disease.[12]

Rule 5: Act and Organize Locally, but Cultivate a Global Vision and Solidarity

Our food, farming, land use, energy, and commerce needs to be not only regenerative, eliminating the use of fossil fuels, toxic chemicals, GMOs, and dangerous drugs, but also relocalized. We need global-scale change, but we need it to unfold locally, in the thousands of towns, cities, and rural counties where people live and work in the United States, as well as in the million towns, cities, and rural communities where the world's population lives. This means getting involved in changing local consciousness, local politics, and local policies as well as making changes on the regional, national, and international levels. If we want to win people over and inspire them with hope, we need to be able to point to positive examples of food, farming, landscape management, resource management, renewable energy, education, and other regenerative practices in their local communities or regions that embody the principles that we're talking about. We're not going to be able to inspire a critical mass of people with just stories about renewable energy, regenerative and organic farms or ranches, or reforestation projects taking place across the continent or across the world. We need to find, publicize, and build up local examples and pilot projects that prove our concepts. And we need to have a plan for local, regional, and national policy change and public/private grassroots financing and technical assistance to scale up these positive examples, rather than counting on philanthropy, international banks, Wall Street, and corporate agribusiness to get the job done. We need to educate public officials (of whom there are more than 500,000 in the United States alone) on the local, state, and federal level so that they understand regenerative principles and are willing to enact regenerative policies and practices, rather than maintaining business as usual—or we have to elect new officials to take their place.

Fifty to sixty years ago, we had a lot less CO_2, methane, and nitrous oxide in the atmosphere, as well as more carbon organic matter, fertility, moisture, and biodiversity in our soils, forests, and landscapes. Most of the foods in our locally owned grocery stores and restaurants were organic by default (produced with a minimum of chemicals, synthetic inputs, and animal drugs), were far more nutritious than comparable foods today, and typically came from within a 100-mile radius of our homes. Most of it was consumed or prepared fresh, rather than being highly processed. People still cooked and shared most of their meals at home, rather than spending half of their food dollars in restaurants, as we do today in the United States. A significant proportion of Americans still lived in rural communities, and millions still made a living from family farms. Factory-scale farms and fast-food restaurants were rare. These traditional, wholesome food and farming systems are not just relics or throwbacks from the past but, rather, inspirational models for how we need to move from degenerative food, farming, and commerce to twenty-first-century regenerative practices.[13]

We need to rebuild healthy, organic, and relocalized systems of food and farming and repair and restore our local environments if human civilization is to survive. To do this will require Regenerators to put a priority on local and regional education, action, and mobilization, in our personal lives and households, as well as in the marketplace and the political arena. At the same time, we have to inject or integrate a national and global perspective into our local grassroots work and community building.

We certainly need to emphasize, over and over again, that we are facing a global state of emergency that requires a World War II–scale campaign of mobilization and cooperation. But the battle against severe climate change, environmental destruction, deteriorating public health, poverty, political corruption, and societal alienation will be fought and won based upon what billions of us—consumers, farmers, landscape managers, public officials, business owners, students, and others—do (or don't do) in our million local communities as part of a global awakening and paradigm shift. We must think, act, and organize locally, while also cultivating a global vision and global solidarity.

Rule 6: Become a Positive Example of Regeneration

The personal is political. People hear not just the overt message of what we say or write but also our subliminal message—that is, our presence, behavior, and attitude. Only by striving to embody the principles of Regeneration (hope, solidarity, creativity, hard work, joy, optimism) in our everyday lives and practices (i.e., our work, food, clothes, lifestyle, and how we treat others and the environment, vote, spend our money, invest our savings, and spend our time) will we be able to inspire those around us. Only by constantly nurturing our personal health, mental and physical, and our convictions, ethical and political, as well as those of the people around us, will we be able to maintain the strength, self-confidence, and optimism we will need to carry out a Regeneration revolution that will take the rest of our lives.

In the 1960s, when I came of age as an activist, we had a saying: "There is only one reason for becoming a revolutionary: because it is the best way to live." I believe this slogan is just as appropriate now as it was then. One of the wonderful things about Regeneration is that it not only is our duty and our potential salvation but can actually become our pleasure as well. As the farmer-poet Wendell Berry once said, "The care of the earth is our most ancient and most worthy and, after all, our most pleasing responsibility. To cherish what remains of it, and to foster its renewal, is our only legitimate hope."[14]

2

———

Regeneration:
The Big Picture

The health of soil, plant, animal, and man is one and indivisible.

Sir Albert Howard

Nature has this incredible ability that when subjected to disruption it actually regenerates itself a degree or two higher than where it was before.

Allan Williams, "This Farm Is Medicine"

All of the terms and abbreviations (carbon, CO_2, CO_2e, GHG, ppm, tons, gigatons) and numbers (millions, billions, trillions) used to describe global warming and carbon sequestration in this book can be confusing. So let's just start off with the big picture and try to make the math and the science as simple as possible.

Humans are emitting the equivalent of ten billion tons of carbon every year through burning fossil fuels and destructive agricultural and land use practices. These ten billion tons are going up into the already carbon-supersaturated atmosphere and into the oceans, heating up the average temperature on the planet and pushing us closer and closer to runaway global warming and climate catastrophe. If we want to avoid climate chaos and survive, we the global grassroots must rise up and put an end to business as usual over the next decade and beyond. We must cut global emissions of carbon and other greenhouse gases (GHG) by approximately five billion tons, or 50 percent, over the next decade. At the same time, we must simultaneously start to draw down billions of tons of atmospheric carbon every year through regenerative food, farming, and land use practices, sequestering it in our soils, forests, and plants, reaching at least five billion tons of global carbon drawdown by 2030. This will bring us to net zero emissions by 2030, which is basically the bottom line for human survival. For the next several decades beyond 2030, we must continue to draw down and sequester in our soils, forest, and plants approximately five to twenty billion tons of carbon per year, while producing little to no fossil fuel emissions whatsoever. This will move us away from our current trajectory of climate catastrophe and put us on the regenerative path of restabilizing the climate and bringing atmospheric CO_2 back to the safe levels that existed before the industrial revolution began around 1750. Different scientists define the "safe" level of CO_2 in the atmosphere as either 350 ppm, 300 ppm, or 280 ppm, but everyone agrees that the most desirable level to avert climate disaster would be 280 ppm.

For reference purposes in this book, 1 ton of solid carbon is equal in weight to 3.67 tons of CO_2 gas, while CO_2e refers to all greenhouse gases including methane and nitrous oxide. CO_2e is often referred to as the "carbon footprint" of burning fuels or releasing greenhouse gases through land disturbance or deforestation. *One* part per million (1 ppm) of CO_2 in the atmosphere is equal to 7.8 *gigatons* (Gt or billion tons) of CO_2 or 2.125 Gt of solid carbon. Global atmospheric concentrations of CO_2 reached 415 ppm in 2018, the highest level in millions of years, whereas safe levels are considered to be 280 to 350 ppm, the range that has existed since humans appeared on the planet.

Regeneration and Carbon Drawdown

If you're not familiar with the amazing potential of regenerative farming and land use to not only improve the quality of our food but to draw down enough carbon from the atmosphere to reverse global warming, you're not alone. One of the best-kept secrets in the world today is that, along with switching to renewable energy, the major solution to global warming and climate change (as well as rural poverty, nutrient-deficient food, deteriorating public health, civil strife, and more) lies in regenerating what's right under our feet and at the tips of our knives and forks.

As most of us now realize, our very survival as a civilization and a species is threatened by a systemic crisis that has degraded climate stability along with every major aspect of modern life. This mega-crisis cannot be resolved by piecemeal reforms or minor adjustments such as simply reducing our current levels of fossil fuel use, global deforestation, soil degradation, and military spending. Either we move beyond merely treating the symptoms of our planetary degeneration and instead build a new system based upon regenerative food, farming, and land use, coupled with renewable energy practices and global cooperation instead of belligerence, or else we will soon (likely within twenty-five years) pass the point of no return.

As a popular slogan of the climate movement proclaims, we need "System Change, Not Climate Change." The systemic repair and regeneration of our common home, including restoration of balance in the carbon and water cycles between the atmosphere and terrestrial ecosystems, must become a global priority for all. Our broken climate, supersaturated atmosphere, depleted soils, and degraded landscapes constitute the number one security threat for all nations. By bringing our soils, plants, forests, water, biodiversity, and animals back to full life and vigor, we will be able to regenerate not only climate stability, but public health as well. In addition, a Regeneration revolution will enable us to revitalize rural and urban economies; alleviate poverty, hunger, and malnutrition; and rekindle a common sense of hope and mission in the global body politic.

However, this long-overdue renaissance and pacification, the centerpiece of which is the decarbonization of the atmosphere and the recarbonization and refertilization of the soil and biota, will require nothing less than a revolution in thought and everyday action by several billion people living and working in a million different local communities across the world: farmers, consumers, workers, businesspeople, politicians, investors, religious communities, students, and activists. Repair and regeneration of Earth will not be possible without a profound shift in public consciousness, political power, public policy, farm practices, business practices, consumer practices, investments, and international relations. In the United States and many other nations, fortunately, there is a growing discussion and debate about mobilizing on a major scale for systemic climate and economic change—that is, enacting a Green New Deal.

Reversing Climate Change: The Photosynthetic Bottom Line

How do we describe the crisis of global warming and severe climate change in such a way that everyday people understand the problem and grasp the solution (renewable energy and regenerative food, farming, and land use) that we're proposing? The bottom line is that, by burning fossil fuels, destructive land use practices, and other behaviors, humans have put too much CO_2 and other greenhouse gases (methane and nitrous oxide) into the atmosphere, while we have removed or oxidized several hundred billion tons of the carbon which once resided in our soils, plants, and forests. These gases are called "greenhouse gases" (GHG) because they effectively function as an atmospheric blanket, turning the atmosphere into a supercharged greenhouse that traps the sun's heat rather than allowing it to radiate back into space. As a result, the planet is heating up.

Earth's normal carbon cycle, or the balance between CO_2 in the atmosphere and carbon in our soils, plants, and forests, involves plants and trees drawing down carbon from the atmosphere via photosynthesis. Unfortunately, because we are supersaturating the atmosphere with CO_2 emissions, and because our destructive food, farming, and forestry practices have degraded a major portion of Earth's landscapes and thus the global capacity of plant photosynthesis, the carbon cycle is imbalanced: plants are not drawing down enough of these CO_2 emissions to cool things off.

Enhancing the photosynthesis end of the carbon cycle (by boosting the health and volume of plants and trees across the world) is the only practical way in which we can draw down a significant amount of the excess greenhouse gases in our atmosphere that are heating up Earth and disrupting our climate. Through photosynthesis, plants and trees utilize solar energy to break down CO_2 from the atmosphere; they release the oxygen and transform the remaining carbon into plant biomass and liquid carbon. The plants transfer a portion of the liquid carbon that they produce into their root systems to feed the soil microorganisms that in turn feed the plants. From the standpoint of drawing down enough CO_2 from the atmosphere and sequestering it in our soils and biota to reverse global warming, qualitatively enhanced photosynthesis is all-important.

The damage caused by global warming and severe climate change compels us to tackle the problem on two main fronts: stop emitting more greenhouse gases and draw down, through enhanced photosynthesis, atmospheric CO_2 concentrations from our current 415 ppm to 280 ppm (the atmospheric levels that existed in 1750, before the industrial revolution took hold). This means drawing down and sequestering in our planet's soils and biomass approximately 135 ppm of atmospheric CO_2, which amounts to 286 billion tons (gigatons) of carbon. At this scale, drawdown will require a major (and global) increase in tree and perennial cover, biodiverse landscapes (rather than monocultures) everywhere from farms and ranches to towns and urban areas, healthier soils (able to retain more moisture and possessing greater fertility), healthier plants, greater biodiversity, increased aboveground biomass, increased belowground carbon in the form of organic matter and biological life, and the transition of most of the world's seventy billion farm animals from intensive-confinement factory farms to holistic grazing and free-range animal husbandry. The systemic regeneration of the planet's biological systems has the incredible power not only to prevent our current stage of climate change from morphing into climate catastrophe but ultimately to suck down 286 Gt of the preexisting or legacy carbon from the atmosphere, thereby reversing global warming and restoring climate stability.

But before we go any further, let's mention two other significant greenhouse gases besides carbon dioxide (which is, all by itself, responsible for three-fourths of all GHG emissions): methane and nitrous oxide. Methane (CH_4) is a powerful greenhouse gas with twenty-five times the warming potential of CO_2 over a hundred-year period. Methane constitutes approximately 14 percent of all greenhouse gases in the atmosphere and breaks down into CO_2 after eight to twelve years. Methane emissions derive primarily from natural gas extraction and fracking, wetlands, rice paddies, manure ponds on factory farms, and noncomposted food and organic waste thrown into landfills or garbage dumps. Nitrous oxide (N_2O), which constitutes 10 percent of all greenhouse gases, is a dangerous, long-lived gas that can persist in the atmosphere for up to two hundred years. Nitrous oxide has up to three hundred times the warming effect of CO_2 over a hundred-year period. A full two-thirds of all N_2O in the atmosphere comes from the nitrogen fertilizer used on industrial farms, which is just one more reason why we must convert the world's farms to organic practices, whereby organic manure, cover crops, and compost are used to maintain and increase soil fertility, instead of chemical fertilizers.[1]

The current excessive, climate-destabilizing levels of CO_2, methane, and nitrous oxide in the atmosphere—the equivalent of almost 880 billion tons of carbon—are like a heavy insulation blanket hovering above

Earth's surface, trapping the sun's heat. As planetary temperatures rise, more water evaporates from oceans, surface water, landmass, and plants, forming excessive water vapor in the atmosphere and contributing to the warming effect in a negative feedback loop. Although the sun on average is heating up every square meter of Earth with 342 watts, at the present time only 339 watts is radiating back into space. The extra 3 watts of energy per square meter is trapped by greenhouse gases, and it is already wreaking havoc, as we can see.[2]

The shroud of greenhouse gases and water vapor in the atmosphere has now raised the average surface temperature on Earth by 1°C (1.8°F), enough to begin to seriously disrupt the climate, melt glaciers and polar ice caps, raise ocean levels, shift ocean currents and rainfall patterns, reduce biodiversity, cause species extinctions, and reduce crop yields on a significant scale. If this destructive release of greenhouse gases continues, and if we don't begin a major drawdown of legacy carbon that's already lodged in the atmosphere, human civilization will likely collapse by midcentury under the impact of rising seas, superstorms, killer droughts and floods, water shortages, crop failures, increased disease and pestilence, massive forced migration, geopolitical instability, and endless war.

Not only has this massive volume of greenhouse gases supersaturated the atmosphere and destabilized the climate by raising the average surface temperature by 1°C (1.8°F), but it has also decreased the fertility and moisture-holding capacity of our soils, disrupted rainfall patterns, and degraded the level of nutrients in our food, making it increasingly difficult for the world's 570 million farmers and farmworkers and three billion rural villagers to make a living off the land, and making it difficult for several billion urban consumers to find or afford healthy, nutritious food.

Our destructive fossil fuel energy, agriculture, and land use system has also deposited far too much carbon in the oceans, which now are overloaded with 38,000 Gt of carbon, a level that is causing the oceans to heat up and become acidic, threatening the survival of fish and marine life. The Earth's oceans absorb CO_2 from contact with the air and form carbonic acid or bicarbonates, lowering the ocean's pH levels and making the water more acidic. This oceanic pool of carbon is increasing by 2.3 billion tons per year because it is still absorbing CO_2 from the atmosphere. For the moment there's not much we can do to reduce the carbon overload in the oceans other than to stop emitting CO_2 into the atmosphere. But drawing down billions of tons of excess carbon from the atmosphere through enhanced photosynthesis into our soils is something we can and must do.

Unfortunately, the extra 286 billion tons of carbon and other greenhouse gases already released into the atmosphere have potentially "baked in" a future temperature increase that will be made catastrophic (unless they're massively drawn down and sequestered in our forests and soils) through feedback mechanisms such as increased ocean temperatures, melting polar ice caps, increased forest fires, and more releases of greenhouse gases from degraded and decarbonized soils, forests, and wetlands.[3] These feedback mechanisms include the phenomena that increased ocean temperatures cause more water evaporation, which causes the atmosphere to hold in more heat. Melting arctic ice causes the polar oceans to absorb more heat instead of reflecting this heat back into space. Melting permafrost releases methane into the atmosphere. Rising temperatures dry out the Earth's forests (as well as wetlands and peatlands) creating the conditions for more forest fires, which then release even more CO_2 into the atmosphere. Cutting down forests, ploughing up agricultural lands, overgrazing, and applying synthetic chemicals to crops that deplete the carbon sequestration potential of plants and soil are also releasing enormous amounts of CO_2, methane, and nitrous oxide into the atmosphere.

According to global scientists the projected extreme rise in global temperatures, if current business as usual continues, will fry the planet, kill all major crops, and put an end to human civilization, making it imperative that we get to net zero emissions and carry out a global transformation of our energy use, agriculture, and land use practices.

The Great Drawdown

The amount of CO_2 we need to draw down into our soils, trees, and landscapes to achieve the prescribed safe balance of 280 ppm of CO_2 in our atmosphere amounts to approximately 286.2 billion tons (gigatons) of stable carbon (1 ppm CO_2 = 2.12 Gt carbon). In other words, we need to draw down enough excess carbon so that we end up with approximately 594 billion tons of carbon (approximately 2.18 trillion tons of CO_2e) in the atmosphere, instead of our current 879 billion tons (approximately 3.23 trillion tons of CO_2e).

A drawdown of 286 Gt of carbon may sound like a lot—in fact, it *is* a lot—but keep in mind that humans have released between 320 and 537 Gt of carbon from our soils and forests up into the atmosphere since the advent of agriculture ten thousand years ago.[4] This carbon release has taken place through deforestation and destructive agricultural practices (plowing, monocropping, erosion, and more recently the use of massive amounts of synthetic fertilizers and pesticides) that destroy the soil, forests, and plants' natural ability to draw down and sequester CO_2. More than half of these agricultural and land use emissions have occurred since 1850.

Our living soils and forests, in fact, based upon real-life farming and land use practices, have the capacity to sequester and safely store enough carbon to bring us back to the safe levels of CO_2 in the atmosphere that we had before the advent of the industrial revolution. Unfortunately, we need to carry out this drawdown rather quickly, over the next twenty-five to thirty years, to avoid triggering the deadly feedback mechanisms mentioned earlier that will threaten to bring on runaway, basically unstoppable, global warming.

Humans have also put another 300 Gt of carbon into the atmosphere from burning fossil fuels (coal, oil, and gas) since 1750. This has to stop. The Paris Climate Agreement calls for net zero emissions by 2050, while the proposed Green New Deal in the United States (see chapter 5) calls for net zero emissions by 2030. In either case, renewable energy, especially solar and wind power, will have to replace fossil fuels as soon as possible. Fortunately, the renewable energy revolution is already moving forward rapidly, driven by innovation, consumer demand, public policy, and business investment.

To meet our goal of 286 Gt, we will have to begin to draw down *annually* 10 Gt of legacy carbon from our supersaturated atmosphere. At the present time our degraded forests, farmlands, and pastures are drawing down only 1.2 Gt of net carbon (meaning they draw down 1.2 Gt more carbon than the amount they release) every year. But they have the capacity to draw down many times more.[5] And if we are able to draw down enough CO_2 from our atmosphere to bring us back to the safe level of 280 ppm, we will end up with approximately 2.75 trillion tons of carbon sequestered in our soils and forests, instead of the current 2.5 trillion tons. We'll also end up with soils that are much more fertile (carbon-rich) and better able to retain moisture, more and healthier forests, and healthier animals and food, among other benefits.

As world-renowned soil scientist Dr. Rattan Lal from Ohio State University has stated, "A mere 2 percent increase in the carbon content of the planet's soils could offset 100 percent of all greenhouse gas emissions going into the atmosphere."[6] And it's doable; some of our best regenerative farmers have increased their farm soil's carbon content by 1,000 percent, from 1 percent to 10 percent.

Many, if not most, of the world's twelve billion acres of croplands and pasturelands and ten billion acres of forests are degraded. Millions of acres are so degraded and eroded that they are becoming desertified and infertile. With degradation comes carbon depletion in the soils and biomass. Scientists estimate that if even half of these carbon-depleted, degraded landscapes could be regenerated and brought back to full life, with maximum biomass growth and carbon sequestration capacity, especially those lands that lie in the tropical and temperate climate zones, we would stop global warming.[7]

Of course, this will require major changes, not only in agriculture and land use practices, but global changes in consumer consciousness, marketplace behavior, political power, public policy, business priorities, and investment practices as well.

Luke Smith, CEO of Terra Genesis International, a regenerative agriculture consulting company, lays out a global regeneration blueprint for five billion acres of the world's degraded lands in a simple formula: "Assuming a conservative average sequestration of 5T/ha/yr it would take just thirty years to return our global climate to below pre-industrial levels."[8] In plain English, that's five tons per hectare (2.47 acres) per year.

Smith's simplified sequestration formula is as follows:

$$5 \text{ tons of carbon per hectare} \times 2 \text{ billion hectares} \times 30 \text{ years}$$
$$= 300 \text{ Gt of carbon}$$
$$300 \text{ Gt of carbon} \div 2.12 = 141.5 \text{ ppm} \ (1 \text{ ppm } CO_2 = 2.12 \text{ Gt carbon})$$

$$415 \text{ ppm} - 141.5 \text{ ppm} = 273.5 \text{ ppm}[9]$$

In short, a Regeneration revolution not only is *absolutely necessary* if we and our children are to survive, but as the formula above indicates, based upon real-world practices and data, it is *absolutely possible*, utilizing already existing, shovel-ready agricultural, forestry, and ecosystem restoration practices to restore Earth's most degraded lands.

Regenerative food, farming, and land use, combined with the transition to 100 percent renewable energy, gives us our best and last chance not only to survive and restabilize the climate but to *thrive*—with healthier food, fiber, animals, people, and local economies as our reward. By bringing together the world's 570 million farmers, farmworkers, ranchers, herders, and fisherfolk with several billion of the world's urban consumers—workers, students, policy makers, businesspeople, and investors—we can safeguard our common home and our common future.

In theoretical terms, as Luke Smith's formula above shows, we can certainly reverse global warming and repair the biological support systems of Earth within three decades. But in practical terms, this great drawdown will require a tremendous change in local-to-global farmer practices, consumer awareness, marketplace behavior, public policy, and economic investment practices. Instead of subsidizing degenerative food, farming, and land use practices to the tune of $500 billion per year, and fossil fuels with $5.3 trillion a year (according to the International Monetary Fund), as the world's governments and financial institutions are doing presently, we will need to force these institutions instead to subsidize the transition to renewable energy and a regenerative food and farming system, especially helping small farmers and ranchers who produce most of the world's food, and to invest several trillion dollars in large-scale ecosystem restoration as well.

At the same time, instead of supersizing themselves on factory-farmed meat, dairy, poultry, and highly processed fast food, a critical mass of working-class and middle-class consumers will have to shift to healthier, more climate-friendly diets, reducing their intake of meat and animal protein from factory farms to zero, and consuming only animal products raised in a humane, regenerative manner.

Regeneration on the global scale that we need will not simply arise out of some polite debate or academic exercise. It will require mass political insurgency and policy change not seen since the wartime mobilization of World War II and the Marshall Plan that rebuilt Europe afterward. It will require hundreds of millions of us to change our political behavior and our purchasing habits, with a cumulative impact that will disrupt and overturn the status quo of the fossil fuel industry, Wall Street finance, industrial food and farming, destructive land use, and other degenerative corporate and government policies. This great transition will require the creation and mobilization of a united front of the global grassroots in all of the 195 nations on Earth, driven by necessity, but inspired by the positive, mutually reinforcing, life-promoting regenerative practices that are starting to spread across the globe.

As noted author and activist Vandana Shiva has pointed out, "Regenerative agriculture provides answers to the soil crisis, the food crisis, the health crisis, the climate crisis, and the crisis of democracy."[10]

Major Drivers of Regeneration and Degeneration

Four major driving forces of Regeneration (or, conversely, Degeneration) are (1) grassroots consciousness (or lack of consciousness), political involvement, and buying power; (2) farmer and land manager innovation (or lack of innovation) and stewardship (or abuse) of the land; (3) enlightened (or misguided) public policy; and (4) regenerative (or degenerative) public and private investment. We will discuss in some detail how we can strengthen and scale up these positive drivers of Regeneration in the following chapters. But first let's briefly review some of the major degenerative forces currently holding us back.

Grassroots Consciousness and Morale

When literally billions of people, a critical mass of the 99 percent, are hungry, malnourished, or struggling to survive with justice and dignity; when the majority of the global body politic are threatened and assaulted by a toxic environment and food system; when hundreds of millions are overwhelmed by economic stress due to low wages and the high cost of living; when hundreds of millions are weakened by chronic health problems or battered by floods, droughts, and weather extremes; when seemingly endless wars and land grabs for water, land, and strategic resources spiral out of control; when indentured politicians, corporations, and the mass media conspire to stamp out participatory democracy in order to force a "business-as-usual" paradigm down our throats, then regenerative change, Big Change, will not come easily.

Disempowered, exploited people, overwhelmed by the challenges of everyday survival, usually don't have the luxury of connecting the dots between the issues that are pressing down on them and focusing on the big picture. It's the job of Regenerators to emphasize the connections between the climate crisis and people's everyday concerns, such as food, health, jobs, and economic justice, and to globalize awareness, political mobilization, and, most of all, hope. As we discussed in the introduction, Regenerators have to be able to make the connections between different issues and concerns, identify and support best practitioners and policies, build synergy between social forces, and effectively lobby governments, businesses, and investors for change, all while educating and organizing grassroots alliances and campaigns across communities, constituencies, and even national borders. But this, of course, will not be easy, nor will it take place overnight.

Our profoundly destructive, climate-destabilizing food and farming system, which is primarily based upon industrial inputs and practices, is held together by a multibillion-dollar system of marketing and advertising that has misled or brainwashed a global army of consumers into believing that cheap, artificially flavored, "fast food" is not only acceptable but normal and natural. After decades of consuming foods overladen with sugar, salt, carbohydrates, and "bad fats" from industrial farms, animal factories, and chemical manufacturing plants, many consumers have literally become addicted to the artificial flavors and aromas that make superprocessed foods so popular.[11]

Degenerate "Conventional" Farms and Farming

Compounding the lack of nutritional education and choice, poverty, inertia, and apathy of a large segment of consumers, other major factors driving our degenerative food and farming system include the routine and deeply institutionalized practices of industrial and chemical-intensive farming and land use (monocropping, heavy plowing, pesticides, chemical fertilizers, GMOs, factory farms, deforestation, wetlands destruction). These practices, which are destructive to the soil, environment, climate, and human health—are especially prevalent on the world's fifty million large farms, which, in part, are kept in place by global government subsidies totaling $500 billion a year. Meanwhile, there are few or no subsidies for organic or regenerative farmers, especially small farmers (80 percent of the world's farmers are small farmers), nor for farmers and ranchers who seek to make the transition to organic or regenerative practices. Reinforcing these multibillion-dollar subsidies for bad farming practices is a global network of agricultural research and teaching institutions, controlled by chemical companies and agribusinesses, that are focused on producing cheap food and fiber (no matter what the cost to the environment, climate, and public health) and agricultural export commodities (often pesticide-intensive GMO grains). Of course, what we need instead are subsidies, research, and technical assistance for farmers and ranchers to produce healthy, organic, and regenerative food for local, regional, and domestic markets, rewarding farmers with a fair price for producing healthy food and being stewards, rather than destroyers, of the environment.

Public Policy and Investments

Agriculture is also the largest employer in the world, with 570 million farmers and farm laborers supporting 3.5 billion people in rural households and communities.[12] In addition to workers on the farm, food-chain workers in processing, distribution, and retail make up hundreds of millions of other jobs in the world, with over 20 million food-chain workers in the United States alone (representing 17.5 percent of the total workforce). This makes public policy relating to food, farming, and land use very important. Unfortunately, thousands of laws and regulations are passed every year, in every country and locality, that basically prop up so-called "conventional" (i.e., industrial, factory farm, export-oriented, GMO) food and farming, while there is very little legislation passed or resources geared toward promoting organic and regenerative food and farming. Trillions of dollars have been, and continue to be, invested in the conventional food and farming sector, including trillions from the savings and pension funds of many conscious consumers who would no doubt prefer to have their savings invested in a different manner, if they knew how to do this. Unfortunately, only a tiny percentage of public or private investment is currently going toward organic, grass-fed, free-range, and other healthy foods produced by small and medium-sized farms and ranches for local and regional consumption.

Monopoly Control

Another driver of degeneration, one that holds back farmer adoption of regenerative practices and determines the type of foods and crops that are produced, is the monopoly or near-monopoly control by giant agribusiness corporations over much of the food system, especially in the industrialized countries, as well as the monopoly or near-monopoly control by giant retail chains such as Walmart and internet giants like Amazon. Author Wenonah Hauter calls this out-of-control system a "foodopoly" (in her book of the same name), and it is designed to maximize short-term profits and exports for the large transnational corporations, preserve patents and monopoly control over seeds, and uphold international trade agreements (NAFTA, WTO) that favor corporate agribusiness and large farms over small farms; factory farms over traditional grazing and animal husbandry, and agro-exports instead of production for local and regional markets.

Food and farming is the largest industry in the world, with consumers spending an estimated $7.5 trillion a year on food. The largely unacknowledged social, environmental, and health costs of the industrial food chain amount to an additional $4.8 trillion a year.[13]

Healthy soil, healthy plants, healthy animals, healthy people, healthy climate, healthy societies . . . our physical and economic health, our very survival as a species, is directly connected to the soil, biodiversity, and the health and fertility of our food and farming systems. Regenerative organic farming and land use can move us back into balance, back to a stable climate and a life-supporting environment.

It's time to move beyond degenerative ethics, farming, land use, energy policies, politics, and economics. It's time to move beyond "too little, too late" mitigation and sustainability strategies. It's time to inspire and mobilize a mighty global army of Regenerators, before it's too late.

3

Grassroots Awareness, Political Mobilization, and Marketplace Demand

< regeneration driver one >

There aren't many families who would have to be convinced to eat locally grown organic health food if it were available and they could afford it. Who wouldn't drive a Tesla, put up solar panels, or buy an energy-efficient home in a walkable neighborhood with great public transportation? Everyone wants these things. We all want to enjoy good health, breathe clean air, and drink pure water. The problem isn't the will. The problem is that the 99% don't have the money to save the planet. We've got student debt. Our mortgages are underwater. We've got medical bills and child care to pay for. And that's if we've been lucky enough to have these opportunities! Many of us have been too poor to go to college, buy a house, or start a family. Our country's struggling family farmers have the same problem. Sure, they'd love to go organic and pay their workers fairly. They want to do what's best for their families, their communities, and their environment. They just have to figure out how to avoid foreclosure and bankruptcy first.

ALEXIS BADEN-MAYER, political director,
Organic Consumers Association, in a December 2018
e-mail urging groups to sign on to the Green New Deal

In 1992 I was hired by the environmental and futurist author Jeremy Rifkin of the Foundation on Economic Trends (FET) to organize a new nationwide consumer movement called the Pure Food

Campaign (PFC). The mission of the PFC was to expose the health, environmental, and ethical hazards of industrial agriculture, factory farms, and GE (genetically engineered) foods, later called GMOs (genetically modified organisms), and to build up mass market demand for organic foods and sustainable agriculture.

I was persuaded by Rifkin's visionary perspective, media skills, and powerful personality to take on the job of FET's national campaign director. This required me to move from Minnesota to Washington, D.C., and to switch my activist focus from anti-war, anti-nuclear, and Central America solidarity work to consumer-oriented food and farm campaigning. Rifkin's prescient writings at the time included his 1989 book *Entropy: Into the Greenhouse World*, the first book on global warming (along with Bill McKibben's *The End of Nature*) written for a mass audience. Reading about the "greenhouse crisis," as Rifkin called it, shook up my intellectual and activist priorities. I began to realize that global warming and climate change were likely to become the most important issues of our time, overshadowing all the causes that I had been working on since the 1960s. In the early 1990s, even though a number of scientists had been sounding the alarm for over a decade, the central importance of global warming was not yet recognized by most people, even the most dedicated activists.[1] In 1992 Rifkin wrote *Biosphere Politics: A New Consciousness for a New Century*, which was basically a manifesto to address the greenhouse crisis. In *Biosphere Politics*, Rifkin outlined the need for a new scientific, ethical, political, and economic paradigm. He called for the building of a US and international movement to reverse global warming, corporate domination, militarism, and societal collapse.

Biosphere Politics, which prefigured the contemporary Regeneration movement by several decades, connected the dots between all the burning issues of the time, including food, agriculture, and mindless consumerism, while providing a radical yet practical roadmap for systemic political change. In retrospect, the book was, in fact, a pioneering call for a Green New Deal. In the same year that *Biosphere Politics* was published, Rifkin made an ambitious, unfortunately abortive attempt to bring together a broad national coalition of all the different green and social change movements in the United States, called the Green Movement for Environmental

Justice and a Sustainable Economy (GMEJ) (I attended the founding meeting as an enthusiastic participant). Unfortunately, leaders of the fragmented activist rainbow, including labor, environment, peace, social justice, and agriculture, were unable to connect the dots between their single issues and different constituencies and work together. The GMEJ movement collapsed shortly after it was convened.

After joining FET, I began organizing, along with my young and enthusiastic staff, campaigns against McDonald's as well as genetically engineered foods and crops, several of which were poised to go on the market. From 1993 to 1994, we managed to organize three thousand protests and picket lines against McDonald's in several hundred cities, handing out over a million leaflets exposing the food giant's degenerative practices, including its reliance on junk food ingredients, factory farms, feedlot beef, and industrial dairies and the manner in which its policies contributed to rainforest destruction and labor exploitation. Despite our efforts, and the efforts of a nationwide network of volunteers whom we recruited, McDonald's refused to budge on our demands to change their menu, labor practices, and supply chain. We obviously needed to build a more powerful movement if we were going to change the behavior of any fast-food giants.

In the wake of the McDonalds's campaign, we kicked off a consumer education and marketplace pressure campaign against the world's first commercial GMO, Monsanto's recombinant bovine growth hormone (rBGH), which was being injected into dairy cows to force them to produce more milk. Monsanto and their Big Pharma and industrial dairy allies forced rBGH onto the market despite damage to cows' health and warnings by scientists of increased antibiotic residues and cancer risks in the milk from cows treated with the genetically engineered hormone. Our anti-Monsanto bovine growth hormone campaign featured "milk dumps" and colorful protests by dairy farmers, chefs, moms, animal welfare activists, and consumer activists out in front of grocery stores and fast-food outlets across the United States. The campaign successfully generated enough marketplace pressure to convince hundreds of dairies to remain rBGH-free and mobilized volunteer grassroots lobbyists at the state and federal levels. A dozen states eventually passed rBGH-free labeling laws, giving dairies the right to label their milk and dairy as rBGH-free. We also coordinated protests and media events with anti-rBGH campaigners in Canada, Japan, Europe, and other countries.

The rBGH campaign ultimately educated and turned millions of consumers against GMOs, in part because our bold, often theatrical tactics and blunt language ("pus milk," "Frankenfoods," "cancer risks") generated major media coverage—much to the surprise and chagrin of Monsanto and the factory-farm dairy lobby. Because of our efforts, rBGH was kept off the market in Europe and Canada and basically marginalized in the United States, with only 10 percent or so of US dairies actually injecting their cows with rBGH on a consistent basis.

Similarly, after protests ("tomato dumps" outside supermarkets), technical glitches (the tomatoes didn't taste that good), and overall bad publicity (including premarket tests that showed damage to lab animals), the genetically engineered Flavr Savr tomato was taken off the market.

Up until the rise of our new anti-GMO, pro-organic movement in the 1990s, utilizing consumer education and boycott power as a major tactic (spearheaded by groups such as the Pure Food Campaign, Greenpeace, Friends of the Earth, Navdanya, and other allies), there had never before been a mass-based consumer movement focused on food and farming in the United States, nor in most of the rest of the world. (Notable exceptions include Gandhi's salt and cotton boycotts of British goods in the India independence movement, and boycotts of German products leading up to World War II.) But as I and other movement activists quickly learned from the anti-GMO and 1998 SOS (Save Organic Standards) campaigns, organized grassroots power can change the marketplace and public policy. Mass education, market pressure, and consumer campaigning had the power to tarnish corporate brands, reduce the market share of industrial agriculture and GMOs, and change government policy. In 1998, with a nationwide mobilization of organic consumers, we stopped the US Department of Agriculture (USDA) from allowing GMOs, nuclear irradiation, and toxic sewage sludge in organics. Just as important, these grassroots forces also had the power to generate billions of dollars in consumer market demand for non-GMO and organic foods and products, not only in the United States, but across the world.

As Rifkin never tired of reminding us, our job as activists was to "change public consciousness" on food, farming, and GMOs. Through public education, media work, coalition building, direct action, litigation, and boycotts, our ongoing mission was to discredit GMOs and factory farms and to mobilize a critical mass of consumers to vote with their pocketbooks for a healthy, organic, and environmentally friendly system of food and farming and to get involved in politics and policy change. This is exactly what I and my counterparts have been doing for the past twenty-five years. And this is exactly what we need to need to do now in order to change public policy and build up market demand for regenerative food and farming—the next soil-building, carbon-sequestering, climate-friendly stage of organics and agroecology.

When I started working with the Foundation on Economic Trends in 1992, the US certified organic market amounted to approximately $1 billion a year. Twenty-five years later, after founding the Organic Consumers Association (OCA), an offshoot of the Pure Food Campaign, our US food and farming movement managed to build up a nationwide network of millions of conscious, anti-GMO, pro-organic, and, more recently, pro-regenerative consumers. Over time this corps of conscious consumers has boosted sales of organic, grass-fed, and non-GMO products to $85 billion a year. Today the OCA and its Mexican sister network, Vía Orgánica, have over two million members and supporters. Our allies have attracted millions of other health-conscious consumers to their subscription lists, customer base, and social media. Along with Dr. Mercola and Dr. Bronner's, two leading organic companies, and a number of others, including Vandana Shiva's India-based Navdanya organization, the Rodale Institute, and leaders from the International Federation of Organic Agriculture Movements (IFOAM-Organics International) network, in 2014 we began recruiting the founding members for a new global movement and network: Regeneration International.

Over the decades our new food movement has been able to educate and mobilize like-minded consumers and farmers all over the world, building up an alternative food and farming network in over a hundred nations. In 1999, at the "Battle of Seattle," our anti-GMO, antiglobalization alliance stopped the World Trade Organization (WTO) from extending its corporate trade rules and monopolistic seed patents over small farmers and world agriculture. At the global climate summits in Paris in 2015, Morocco in 2016, Germany in 2017, Poland in 2018, and Madrid in 2019, our international networks, now focused on regenerative food and farming, delivered our message: "Cool the Planet, Feed the People."

In the period between 2012 and 2015, the US anti-GMO/pro-organic food movement, led by the OCA, Mercola.com, Dr. Bronner's, a dozen state GMO-free coalitions, and others, managed to mobilize millions of voters, raise millions of dollars, and put mandatory bans or GMO food-labeling ballot initiatives on the agenda in numerous states, cities, and counties, heightening the debate on GMOs, industrial agriculture, and the future of farming. Although Congress and the Obama administration managed to kill mandatory GMO food labeling in the summer of 2016 with the passage of the so-called "DARK Act" (the Deny Americans the Right to Know Act), millions of conscious consumers have become aware of the health and environmental hazards of GMOs, factory farms, and pesticides such as Monsanto's Roundup (and its active ingredient, glyphosate).[2] Food safety, diet-related diseases, children's health, factory farms, exploitation of food workers, lack of transparency in food labeling—all have become major concerns. Millions of conscious consumers have begun to vote with their pocketbooks, in ever increasing numbers, for organic and now "beyond organic" regenerative foods and products. In fact, recent polls show that the majority of American consumers buy organic products either always or at least occasionally.

After several decades of pressure from consumer activists and a seemingly unending stream of food safety scandals, Big Food has continued to lose credibility and market share. Aided and abetted by corrupt politicians and powerful trade organizations such as the Grocery Manufacturers Association (now disbanded), the majority of large food corporations alienated millions of consumers by fighting against mandatory "country of origin" and GMO labeling of foods. Watching consumers turn away from their products, large multinational food and beverage corporations such as General Mills, Nestle, Campbell's, Coca-Cola, Cargill, Pepsi, Kellogg's, Danone, Perdue, Unilever, and others have been forced to try to shore up their reputations and market share by buying up every sizable organic brand willing to sell out.[3]

At the same time, giant supermarket chains in North America and across the world, including Walmart, Kroger, Safeway, and Amazon/Whole Foods, have been forced to increase the sales and marketing of their store-brand private-label organic and "natural" products. Even fast-food chains such as McDonald's, Burger King, and Subway have been forced to expand their menus and put more emphasis on nutrition, pressured by lackluster sales among millennials and competition from natural/non-GMO food upstarts like Chipotle. Unable to shore up their sagging profits with organic acquisitions alone, the food giants have hired an army of PR firms and political lobbyists to help them fraudulently "greenwash" and market billions of dollars of their conventional (produced by chemical-intensive or factory farming) products as "natural," "all-natural" or "eco-friendly." In response, groups including the Organic Consumers Association, Beyond Pesticides, and Friends of the Earth have launched numerous lawsuits, suing companies for fraudulently labeling their products as natural, pasture-raised, eco-friendly, or US-made when in fact they are not.

Despite all their efforts, Big Food still finds itself on the defensive, desperately trying to reach out to ever more conscious and savvy consumers and counteract what we food activists have now been telling consumers for twenty-five years: industrial, GMO-tainted, pesticide-laden, factory-farmed foods are bad for your health, bad for farm animals, bad for small farmers and farmworkers, bad for the environment, and, as we're starting to understand, bad for the climate.

Needless to say, all of us in the organic and regenerative food and farming movement now understand that educating and organizing the global grassroots, especially the world's several billion concerned consumers, is perhaps our most important task. There will be no Regeneration revolution in food and farming without a profound transformation in grassroots awareness and fundamental policy change in regard to food, farming, and land use. These changes in politics and public policy, in turn, must be amplified by the economic impact of transforming consumer consciousness and buying habits. Marketplace pressure, especially when scaled up internationally, has the power to force even the largest corporations to change their practices. Uniting conscious consumers and conscientious farmers all around the world against factory farms, industrial agriculture, GMOs, and junk food and in support of organic and regenerative food and farming is not just a lifestyle and consumer choice but a matter of life or death.

The Degeneration of
Food Culture and Home Economics

The culture and nutritional content of food in the United States and many other nations has seriously deteriorated, with a corresponding decline in public health and the environment. There are more than 200,000 fast-food restaurants in the United States alone, serving fifty million customers daily, with popular menu items such as factory-farmed meat, dairy, and poultry and fried potatoes, all of which are high in fat, salt, sugar, and carbs and low in nutrition. Seventy percent of what Americans eat every day is highly processed, essentially junk food, supplying the majority of the average person's calories.[4] This low-grade convenience food, which author Michael Pollan has aptly described as a "food-like substance," rather than real food, is routinely tainted with pesticide and drug residues and laced with a variety of five thousand different preservatives, flavor enhancers, colorings, and other additives, most of which, singly or in combination, have never been tested for human safety.[5]

The world's junk food comes from a supply chain of highly industrialized farms, concentrated animal feeding operations (CAFOs), and food processing plants. This junk food is packaged, distributed, discarded, and wasted (more than 30 percent of food in the United States is thrown away and tossed into landfills, rather than properly composted, while 30 percent of crops are left to rot in the fields)[6] in an energy-intensive, chemical-intensive, climate-destabilizing manner. Diet-related diseases, tragically even among children, are now approaching epidemic levels in the United States as well as many other nations. There are now more mal-fed, overweight, and obese people in the world (1.9 billion), many of whom have been supersized by a cheap junk food diet that is low in nutrition but high in carbs, sugar, and bad fats, than there are malnourished people suffering from hunger (870 million).[7]

Even in the United States, by many standards the richest nation in the world, many consumers say they'd like to purchase more organic food, fresh local vegetables produced without chemicals, and grass-fed meat and dairy, but they can't afford to. The main reason for this is *not* that organic and regenerative foods are priced too high in terms of what organic farmers are paid, their production costs, or their actual value and nutrition.

Most organic farmers and farm workers in fact deserve and need *more* money for their products than they are currently receiving. Rather, organic and regenerative food *seems* too expensive to many consumers because the combined high costs of modern living (for housing, transportation, insurance, health care, education, credit card debt) make nutrition-dense, healthy food seem like an unaffordable luxury. Meanwhile, the enormous hidden costs of cheap industrial food (damage to public health, the environment, and the climate) are routinely ignored or underreported. Of course, government policies distort the real cost of food by subsidizing industrial agriculture and highly processed foods, making local and organic foods seem more expensive.

In 1950s, the average US household spent about 29.7 percent of its income on food (much of it local, fresh, and produced naturally), and people weren't typically complaining about high food costs.[8] In the United States today, the average household spends only about 10 percent of its income on food (half of which is spent eating out in restaurants instead of cooking at home). Yet many Americans claim that they buy highly processed, industrially produced food because it's cheaper and that's all they can afford. This is understandable for lower-income households (the bottom 20 percent of whom have a median household income of only $11,000), since they are spending proportionately 50 percent more of their household income on food. (This is why food stamps, especially food stamps geared toward healthy food, are extremely important.) People today in Europe spend far more than Americans on food as a proportion of their household income. For example, in Italy, France, and Spain, consumers spend 50 percent or more (15 percent or more of their household income) on food than American consumers, and consider this "normal."[9] (Then again, in most European countries health care and education are free and mass transportation is affordable, so consumers have more disposable income.)

In order to move forward on regenerative food and farming, consumers will need to be educated on the real costs of food and understand that organic and regeneratively raised food cooked at home is far cheaper than undermining our health and destroying the environment. Real fresh food, cooked at home from scratch, is well worth the time and investment in terms of improving our health, bringing families and communities together at the table, preserving the environment, and resta-bilizing the climate.

Part of the problem in our fast-food nation of course, is the fast pace of living, time spent commuting to work and school, or, for many, time spent working multiple jobs. Our lack of time is compounded by the fact that cooking skills have declined considerably, in part because home economics is no longer taught in schools, nor considered a priority. An unfortunate trend reinforcing culinary illiteracy is that families and households no longer consider it important to sit down together at home for the evening meal or to invite friends over for a home-cooked meal. The average American now spends ten hours a day in front of computer, TV, and cell phone screens, and very little time in the kitchen or seated together at the family dinner table. There's no way to change American eating habits or our food and farming system without thoroughly educat-ing consumers, including youth, on the importance of food and nutrition, investing in quality organic and regenerative food, and learning how to cook at home because these are the cornerstones of a healthy life, culture, environment, and climate.

The good news is that the culture of food is already changing, with all sectors of the population expressing a noticeable increase in interest in food, cooking, gardening, local food hubs and farmers markets, and nutrition. Gardening, including organic gardening, is now the number one hobby in America, with a full 35 percent of households, especially younger households, getting involved, often growing at least some of their own food.[10] This growing interest in food, gardening, natural health, and healthy living, combined with changes in public policy and the market-place, indicates that spending money for good food, and spending time in the kitchen, will soon once again become the norm in the United States and elsewhere, as it should be.

The additional good news is that most of the world's several billion farmers and rural villagers, especially in the Global South (Asia, Africa, and Latin America), are still small farmers. These farm-ers are often genuinely interested in improving their management of soil fertility, crop nutrition, and animal husbandry. According to the Food and Agriculture Organization (FAO) of the United Nations, 5 to 10 percent of the world's farmers "are already using regenerative, climate-friendly techniques."[11] Most of the food producers in the world (mainly in the Global South, as opposed to the industrialized Global North) are not yet caught up in the trap of chemical-intensive industrial agriculture. In fact, millions of the world's smallholders (small, traditional farmers) are still farming traditionally and sustainably, using few or no chemicals and raising their livestock in a natural manner. If these small farmers can see that expensive chemical inputs and GMO seeds are not necessary to grow nutritious food, if consumers will pay them a fair price for traditional and natural food produced in a healthy agroecological manner, and if the farmers can get help and support (instead of resistance) from their governments, most of them will be willing to maintain and improve upon their agroecological, traditional practices.

The world's smallholders, who feed 70 percent of the world's people with 25 percent of the resources, can become major drivers of Regeneration.[12] The traditional, basically nonchemical, potentially organic and regenerative produce, grain, and livestock produced by several hundred million smallholders for their own and their community's consumption probably amount to a trillion dollars a year across the world today. A major proportion of this production is not counted in marketplace statistics since much of it is grown for the farmers' own consumption or the very local market. But according to the ETC Group, an estimated "70% of the world's population—4.5–5.5 billion people of the world's 7.5 billion people—depend on the Peasant Food Web for most or all of their food."[13] Supporting the world's smallholders, providing them with increased markets, technical assistance, and financing, will be one of the primary tasks of a new global Regeneration movement.

In addition to smallholders in the Global South, in urban and suburban areas around the world, there are approximately a billion gardeners and urban farmers producing at least some their own food.[14] In the Global North, millions of farmers and ranchers are already utilizing organic and regenerative techniques. Millions more are utilizing agroecological practices such as no-till farming, cover crops, increased crop rotation, multispecies and perennial production, and rotational grazing. In addition, a whole new generation of young farmers and ranchers, or would-be farmers and ranchers, is ready to make a change.

The paradigm shift we need in consumer consciousness (a shift that has already begun) will be a major factor in getting a critical mass of farmers, food companies, policy makers, and investors to move away from chemical- and energy-intensive agriculture and adopt or support regenerative practices. But this shift will not just occur spontaneously as individual consumers look around and see what the degenerative practices and policies in the food and farming sector are doing to their health, the environment, and the climate. If real change is to take place, we will need not just a change in the purchasing and consumption patterns of individuals but a change in our entire food culture. And changing our food culture will require an organized Regeneration movement, thousands and eventually millions of volunteer educators and citizen lobbyists, in all of the million cities, towns, and rural communities across the globe.

Five Steps to Becoming a Grassroots Mobilizer

The following five major steps to becoming a Regenerator educator and mobilizer are designed to initiate this global consumer education process, starting at the local level, but eventually spreading into thousands of communities across the globe.

Step 1: Become a Regeneration Educator

Educate yourself on the basic principles of regenerative food, farming, and land use. We must learn how to explain in everyday language what the problem is and what the solutions are. We must deliver the positive message that people all over the world are starting to embrace Regeneration as a guiding principle for food, farming, and land use because it offers a fundamental solution to climate change, deteriorating public health, environmental degradation, and related problems. Once people understand the basic concepts of climate change (most people already do), we must move beyond the prevalent gloom-and-doom talk on global warming and talk about positive solutions that everyone— consumers, farmers, gardeners, landscape managers, educators, students, businesspeople, investors, policy makers, and the entire global body politic—can begin to implement in their everyday lives and practices.

We must develop and hone our understanding and enthusiasm to the point that we can begin to successfully inspire and recruit others to our cause. This may take a while, but through practice we can improve our outreach, communication, and recruiting skills. We can begin by starting conversations with people whom we feel comfortable talking to: people who are concerned about the climate crisis and related issues but haven't yet heard, or haven't heard much, about Regeneration. Talk to those you know and trust before you try to speak to community, school, business, activist, or church groups. Avoid, for the moment, wasting your time arguing with climate deniers and other dogmatists, and strive to reach out to those with an open mind. You'll know you're ready to go out into the community and educate once you can convince and inspire people—one-on-one—in your local circle of family, friends, coworkers, and acquaintances. On the Regeneration International website, you will find a set of basic educational tools—articles, videos, and a PowerPoint presentation—that you can study and then share with your first circle of potential Regenerators. You can also find on that website an annotated bibliography and a daily newsfeed.

Step 2: Form a Core Group

After you've mastered the basics, form or join a small core group with four or five (or more) other people who understand and are truly inspired by the basic principles of regenerative food, farming, and land use. Prime candidates for recruitment might be family members and coworkers; local food, church, climate, political, or farm activists; concerned parents; schoolteachers or students; gardeners and farmers; and artists. Arrange a series of potlucks or study groups to increase your core group's understanding of the issues and to brainstorm about which local groups you could possibly reach out to in order to expand your circle. One of the best local groups to connect with is the youth-led Sunrise Movement. Use the website and social media page of Regeneration International to keep abreast of current developments in the global Regeneration movement. Once you've formed a core group, register your contact information and sign up as an affiliate with Regeneration International.

Step 3: Think and Link Up Globally

Familiarize yourself and your core group with the "4 per 1000" Initiative, a local to international pledge to take action, signed by hundreds of grassroots organizations, cities, states, and nations as part of the Lima-to-Paris Action Agreement arising from the twentieth Conference of the Parties (COP20)—the United Nations' annual climate change summit—to sequester carbon through regenerative practices on a scale that can begin to reverse global warming.

As André Leu, former president of IFOAM (International Federation of Organic Agriculture Movements), explained at the historic Paris climate summit in 2015:

> The French Government's 4 per 1000 Initiative is a fantastic win, win, win for the planet. By changing agriculture to one that regenerates soil organic carbon we not only reverse climate change, we can improve farm yields, increase water-holding capacity and drought resilience, reduce the use of toxic agrochemicals, improve farm profitability and produce higher-quality food. This is the most exciting news to come out of the Paris Climate Summit. . . . This initiative is historic, marking the first time that international climate negotiators and stakeholders have recognized the strategic imperative of transforming and regenerating our global food and farming system in order to reverse global warming.[15]

The importance of the "4/1000" Initiative, as it's often written, is that it is the only global agreement to sequester excess carbon from the atmosphere in order to reverse climate change. Think of the 4/1000 Initiative as sort of a global Declaration of Interdependence, an acknowledgment (and a pledge to take action) coming from a visionary segment of the world's seven billion people to regenerate Earth and mobilize the global grassroots.

Regeneration activists in over three dozen countries are now using the 4/1000 Initiative as a tool to do outreach, to enroll organizations to formally join the Regeneration movement, and to build up coalitions to lobby town, city, county, state, and national governments to pass resolutions and ordinances in support of the initiative. Regeneration International's goal is to get thousands of community-based organizations and NGOs (nongovernmental organizations) to sign on to the 4/1000 Initiative, and then to use this grassroots power to convince thousands of cities, states, and nations of the world to do the same. You can sign up your own group on the 4/1000 Initiative website and join the global movement.

Step 4: Develop an Outreach Plan

With your core group and allies, develop a strategy and a plan of action to reach out, one by one, to as many groups and organizations as possible in your local area and region. Your goal should be to map out and recruit key individuals in key groups, winning them over so that they "connect the dots" between what their organizations are already doing and the global campaign to regenerate the Earth. Target groups for discussion and recruitment should include student groups and teachers; church groups; food, farm, climate, peace, hunger, immigration, and environmental groups; elected public officials and candidates for office; and any other civic organizations with open-minded members. In the United States, it's especially important to reach out to the hundreds of local, state, and national organizations who have registered with the Sunrise Movement in support of the Green New Deal, but who may not yet understand (or fully understand) how regenerative food, farming, and land use can help us achieve our goal of net zero fossil fuel emissions by 2030.

Step 5: Scale Up

Once your local Regeneration core group has carried out significant public education in your area, built up a critical mass of organizations that have formally signed on to the 4/1000 Initiative, and begun to lobby your local town, city, state, or regional governmental bodies to sign on to the initiative, you can contact the Regeneration International office for further advice on how to arrange regional and national meetings, spread the Regeneration message even further, and publicize and scale up regenerative pilot projects and best practices in your region. Be sure to link up with the Sunrise Movement and local supporters of the Green New Deal.

4

Carbon Farming, Reforestation, and Ecosystem Restoration

< regeneration driver two >

If land that is bare, degraded, tilled, or monocropped can be restored to a healthy condition, with properly functioning carbon, water, mineral, and nutrient cycles, and covered year-round with a diversity of green plants with deep roots, then the added amount of atmospheric CO_2 that can be stored in the soil is potentially high.

COURTNEY WHITE, *Grass, Soil, Hope*

Global reforestation could capture 25 percent of global annual carbon emissions and create wealth in the global south.

UNITED NATIONS ENVIRONMENT PROGRAMME, in its website's description of the Trillion Tree Campaign

The very good news we need to deliver is that carbon farming, reforestation, and ecosystem restoration have the capacity, in conjunction with a shift to 100 percent renewable energy, to actually reverse global warming and restore climate stability. Scaled up globally, they have the power to draw down the 286 billion tons of excess, climate-destabilizing atmospheric carbon that were once safely stored underground and put these greenhouse gases back where they belong—in our soils, forests, and landscapes, thereby reversing climate change, improving soil fertility, generating predictable rainfall, enhancing water retention in soils, qualitatively improving food quality, and creating prosperity for the world's three billion farmers, farm workers, and rural villagers.

However, we need to stress that recarbonizing and revitalizing the Earth's soils enough to stop and then reverse global warming requires that we restore and regenerate billions of acres of degraded croplands, pasture, rangelands, forests, wetlands, and peat bogs. According to agronomists and soil experts, between 4.7 and 8.9 billion acres of global farmland, pasture, and rangelands are degraded, including 70 percent of what were once healthy and carbon-sequestering grasslands.[1]

A major portion of these billions of acres, in arid and semi-arid areas of the world, are at serious risk of becoming desertified, unable to support crops or animals, and unable to sequester carbon. Regenerating and restoring these degraded farmlands and soils through organic farming methods, agroforestry, and holistic rotational grazing must become a top priority.

In addition, drawing down 286 gigatons of atmospheric carbon over the next twenty-five to thirty years (the window of opportunity we have left) will require us to preserve, reforest, and expand a significant percentage of the world's current ten billion acres of forest, wetlands, and peat bogs on Earth. Global reforestation on the scale required will also involve planting billions of species-appropriate trees in our urban areas as well. More than 13.6 billion trees have already been planted as part of the United Nations' Trillion Tree Campaign.

Regenerative Agriculture and Carbon Farming

Carbon farming is the process of drawing down atmospheric carbon and sequestering it in the soil by cultivating enhanced soil fertility and plant photosynthesis, basically by managing our carbon, nitrogen, methane, mineral, and water cycles properly. Carbon farming is not just an idea or an untested hypothesis, like most geo-engineering schemes, such as carbon capture and sequestration (CCS) from coal plants. Utilizing the natural ability of plants and trees to draw down CO_2 from the atmosphere and store it in the soil, carbon farming techniques are already being put into practice by millions of farmers and ranchers on hundreds of millions of acres across the world. The stepped-up photosynthesis of carbon farming and regenerative land restoration not only reduces the legacy load of greenhouse gases in the atmosphere but also generates a variety of other benefits that conservationists call "ecosystem services." Ecosystem services include the increased production of soil organic matter, soil carbon, soil nitrogen, and soil microorganisms, leading to increased fertility and higher-quality food, feed, forage, and fiber. Soil organic carbon in the first one to three feet of topsoil can range all the way from less than 1 percent in poor soils to more than 10 percent in optimally regenerated soils.

Regenerative or carbon farming practices also generate greater biodiversity (the density and variety of biological life, both below- and aboveground), including wildlife, and greater water retention capacity. (Rain infiltrates sponge-like healthy soil and is stored belowground, while rain in poor or degenerated soils runs off, eroding topsoil in the process.) Healthy soil also contributes to a farm's resilience to adverse weather (droughts or flooding). Since carbon farming typically is designed to produce a variety of foods or crops (for humans, animals, and the soil) in a particular area, it is important to take note of the enhanced "nutrition per acre" capabilities of regenerative agriculture, in comparison to the "bushels per acre" mantra of chemical farms, GMOs, and commodity crops.

Another way to describe carbon farming is to say that it is the next and "higher" stage of organic farming, livestock management, and land conservation, building upon the healthy and sustainable aspects of previous practices, but moving further, with ongoing qualitative improvement of soil health, pasture health, animal health, food quality, rainfall retention, water quality and availability, and other ecosystem services. Carbon farming is certainly not new, for many of its processes are rooted in indigenous practices that go back for millennia, but many of its techniques have been refined and improved in accordance with the needs of our twenty-first-century environment and marketplace.

Key Techniques of Carbon Farming

A dozen key techniques (there are many more) of carbon farming are as follows:

1. Feeding the soil and increasing soil organic matter with compost, compost tea (a liquid extract of compost), and "green manure" (cover crops)
2. Keeping the soil covered at all times with vegetation, cover crops (especially multispecies cover crops), or mulch and eliminating or reducing plowing and tillage

3. Adding trace minerals or nutrients (from materials such as kelp seaweed or rock dust) that are deficient in specific soils

4. Utilizing biochar (a special type of charcoal) as a soil or compost amendment to enhance the soil's water retention, microbial biodiversity, and fertility

5. Diversifying or rotating crops, rather than growing the same crop on the same land every year (monocropping)

6. Restoring deep-rooted perennial native plants and grasses on croplands and pasturelands, and sowing multiple species of grasses in pasturelands

7. "Alley cropping"—that is, combining the cultivation of rows of annual crops (ideally rotated annually) with rows of perennials

8. Constructing raised beds, terraced rows, berms, and swales that follow the natural contours and water flow of the landscape, and utilizing keyline designs and plows to follow these natural contours to sow crops (especially perennials) along these contoured rows

9. Capturing and storing rainwater in ponds and cisterns and slowing down (though not completely damming) stream flow with constructed barriers

10. Utilizing agroforestry and silvopasture (integrating trees, woody perennials, and succulents into cropland and pastureland)

11. Practicing planned rotational grazing—avoiding either overgrazing or undergrazing by moving herds of livestock in a planned manner across paddocks or pastures

12. Practicing multispecies grazing—grazing different species of livestock on the same pasture, either simultaneously or in succession

Living soils need food, minerals, and nutrition, along with water and sunlight, just like animals and humans. Feeding the soil with organic compost (typically a mixture of biomass—leaves, grass, wood chips, food waste—and animal manure), compost tea (a liquid extract of compost sprayed on plants and soil), and/or cover crops (sometimes known as "green manure"), instead of synthetic fertilizers and chemicals, is the key to restoring and maintaining soil health and enhancing photosynthesis. Dousing the soil with chemical fertilizers, using toxic pesticides, and tilling and compacting the earth with tractors and conventional plows basically kills the soil's biology (especially mycorrhizal fungi) and its natural ability to sequester carbon and to allow rainwater to infiltrate belowground.

Industrial agriculture's chemical fertilizers and pesticides not only kill soil life but are also major drivers of environmental pollution and food contamination. Fertilizer pollution gives rise to phosphorus-clogged, algae-choked waterways and oceanic dead zones, contaminated groundwater and wells, and greenhouse gas emissions. Nitrous oxide emissions, two-thirds of which come from industrial agriculture's heavy use of nitrogen fertilizer, constitute up 10 percent of all climate-destabilizing greenhouse gases.[2]

Cover crops, no-till farming, and crop rotation methods are increasingly being utilized by farmers, even nonorganic farmers growing commodity grains and crops, because these techniques help maintain soil fertility, reduce weeds, and eliminate the need for expensive chemical fertilizers and pesticides.[3] They also increase overall yields and prevent soil erosion. One of the main precepts of regenerative agriculture or carbon farming is to avoid leaving the soil bare and to minimize soil disturbance. This requires the gardener or farmer to forgo plowing entirely, or to use only minimum tillage with techniques such as "double digging" in garden beds or using equipment such as the keyline plow in fields.[4] Carbon farming requires that you keep your garden beds, croplands, and pasturelands covered at all times with perennials, cover crops, or mulch, interspersed wherever possible with food- or forage-bearing trees and bushes, including fencerows, alley crop border perennials, and hedgerows.[5] These regenerative practices all help feed the soil, keep the ground insulated, hold in moisture, prevent wind and rain erosion, fix nitrogen in the soil, and reduce the amount of soil carbon, methane, and nitrous oxide that is released into the atmosphere.

Biochar

Biochar is a special type of charcoal produced by pyrolyzing wood or biomass—that is, super-heating it in a low-oxygen environment.[6] Biochar added to compost and then applied to the soil, or sometimes added directly to the soil, creates an ideal habitat for soil microorganisms (like burying a coral reef in the soil), thereby increasing the soil's biological activity and fertility. Biochar pyrolysis basically allows the combusted material (the charcoal) to retain its microscopic cellular structure, providing trillions of passageways and protected cavities for soil microorganisms to live, and enabling the amended soil to better infiltrate rainfall and maintain moisture. In addition, biochar is very long-lasting in the soil.

By pyrolizing wood or biomass (wood chips, crop residues, manure, brush cuttings, et cetera) instead of burning it all the way down to ash or letting it decompose, half or more of the biomass CO_2 that would have gone up into the atmosphere after the burning or decomposition ends up being safely stored for centuries in the ground.

Plant Diversification and Biodiversity

By planting, rotating, and interspersing a variety of crops, especially deep-rooted native crops and perennials, instead of replanting the same annual crop every year on the same piece of ground, carbon farmers avoid unnecessarily depleting soil nutrients. Carbon farmers typically rotate crops that draw up and deplete nitrogen from the soil (like corn) with legume crops, like beans or alfalfa, which draw down or fix nitrogen from the air into the soil. Crop diversity and rotation also help maintain soil structure, prevent erosion, avoid the buildup of specific plant pests and diseases, and in general preserve and enhance overall biological activity, photosynthesis, and carbon sequestration.

Native plant varieties in general (as opposed to hybrid or GMO commercial varieties) are typically best adapted to the soils, climate variations, and environment of their native area, and they often have deeper roots, temperature tolerance, pest resistance, and other evolutionary adaptations that are absent in commercial hybrids or GMOs, which are designed to grow relatively well anywhere, given ideal conditions and often considerable chemical support from fertilizers and pesticides.[7] Even farms in drier areas that don't have wells for irrigation, which is, in fact, the situation faced by the majority of the world's farmers, can be successful when they utilize native open-pollinated seeds adapted to local rainfall conditions—plants like desert succulents (agave, cactus, et cetera) that draw most of the moisture they need from the air—as well as agroforestry, silvopasture, and other regenerative, water-conserving and water catchment methods.

For each 1 percent increase in soil organic carbon, an acre of farmland or pastureland can store an additional twenty thousand gallons of rainwater in the soil.[8] This enables crops to grow even during periods of low rainfall or drought. Carbon farmers—some of whom refer to themselves as *permaculture farmers*—are able to grow a variety of foods and crops even in semi-arid or arid conditions, in tropical, temperate, and cold weather zones, on either flat or sloping terrain, with limited or no irrigation through the use of contoured terraces, berms (raised beds or rows), swales (canals or ditches), rainwater catchment from the roofs of buildings, rainwater containment ponds, cisterns, and drip irrigation tubing, complemented with strategic planting of woody perennials, desert plants (if appropriate), and trees.[9]

Terraced rows, berms, and swales, laid out in relation to the direction in which water flows when it rains, will slow rainwater runoff and percolate it into the topsoil, while excess water flow can be directed into ponds or cisterns and stored there for livestock or crops in dry periods. Planned rotational grazing of appropriate livestock, both herbivores (cows, sheep, and goats) and omnivores (poultry and pigs), utilizing solar electric or more permanent fencing, can complement the diverse plant and crop production of a carbon farm.[10] Even farms or ranches that normally enjoy adequate rainfall or irrigation can benefit from all of these carbon farming practices.

Carbon Ranching

Dr. Richard Teague of Texas A&M University, a founding member of Regeneration International, explained the principles of planned rotational ("mob") grazing to the US House of Representatives Committee on Natural Resources on June 25, 2014:

> The key to sustaining and regenerating ecosystem function in rangelands is actively managing for reduction of bare ground, promoting the most beneficial and productive plants by grazing moderately over the whole landscape, and providing adequate recovery to grazed plants.[11]

Three-quarters of the world's farmland (twelve billion acres) is pastureland or rangeland (eight billion acres), as opposed to cropland (four billion acres). Billions of acres of these lands are so degraded or desertified that they can barely support grazing, or any agricultural practices, for that matter. The overwhelming majority of pastureland or rangeland has been degraded and decarbonized over time due to either overgrazing or undergrazing.[12]

Livestock pastures, grasslands, or rangelands, contrary to popular belief, need a certain amount of periodic grazing to optimize photosynthesis, to spread manure and urine, to reduce or eliminate dead grasses and invasive species, and in general to carry out regenerative land disturbance. By "regenerative land disturbance," I refer to grazing livestock's natural ability to crimp and crush plant residues, scatter and bury seeds, fertilize the soil with their urine and feces, and create mini-reservoirs for rainfall infiltration and seed germination with their hoof imprints. Unfortunately, several billion acres of valuable grasslands, once covered with deep-rooted, carbon-sequestering native grasses (with roots going down as far as fifteen feet), grazed sustainably and regeneratively over centuries by large herds of migratory herbivores such as bison on the Great Plains of the United States, caribou in the Arctic, or wildebeests of Africa, have been plowed up and planted in shallow-rooted, monoculture grains and row crops, such as corn, wheat, and soybeans. This has significantly reduced global rangeland's ability to sequester carbon.[13]

Carbon farming involving livestock, sometimes called carbon ranching, is basically the practice of planned rotational grazing, controlling the timing, frequency and intensity of grazing, avoiding either overgrazing or undergrazing pastures, moving herds of livestock in an organized manner across the landscape, monitoring their feeding, and controlling their movements, taking into account available grass and forage across delineated paddocks or pastures. Plant photosynthesis, in this case pasture grass or forage, is enhanced or supercharged when an animal bites off the sweeter, most nutritious top one-third of the plant (especially a native plant with deep roots) and then moves on to graze on the tops of other plants in the pasture, rather than grazing the plants all the way to the ground and pulling them up by the roots, which they will do if they are not moved in a timely manner to an adjoining paddock with full-grown grasses.

Techniques of planned rotational grazing include dividing pasture areas into smaller, fenced-in paddocks or corrals, using either permanent fencing or movable solar electric fencing (although shepherds with trained dogs can rotate sheep and other livestock properly without fencing), and "mobbing" or aggregating animals together into a maximum-sized herd for intensive grazing over short periods of time.[14] Related techniques include sowing multiple species of grasses in pastures; grazing multiple species of livestock on the same pasture, either simultaneously or in succession;[15] and silvopasture, which integrates trees and woody perennials into pasturelands. Studies around the world have shown that silvopasture can significantly increase carbon sequestration in pasture soils, even without counting the substantial amount of carbon stored in the aboveground trunk and branches of the tree, succulents such as agaves, and other woody perennial ground cover.[16]

In the United States, Gabe Brown in North Dakota and Will Harris in Georgia are practicing multispecies, multipaddock livestock grazing and sequestering large amounts of carbon in their farms' soil. By combining multiple cover crops, pasture cropping (planting the annual crop directly into the cover crop without plowing, utilizing a seed drill), and multispecies grazing, Brown has seen his soil organic matter rise from 4.2 percent to 11.1 percent from 2006 to 2013, adding up to approximately seven tons of carbon (thirty tons of CO_2e) sequestered per acre per year, with only sixteen inches of rainfall a year and a five-month growing season.[17] Harris has seen similar increases on his farm, where soil carbon levels have increased tenfold, from 0.5 percent to 5 percent or more.[18]

The herds of livestock on carbon ranches do not return to a previously grazed paddock until the pasture has fully recovered. If rainfall or climate conditions reduce the amount of grass and forage available, then the numbers of cattle or durations in paddocks are adjusted. In this way, pasture photosynthesis and carbon sequestration are maximized, without the use of chemical fertilizers or herbicides. Some intensive silvopasture livestock operations, counting the carbon stored aboveground in the trees can sequester between six and fourteen tons of carbon (twenty-two to fifty-one tons of CO_2e) per acre per year.[19] This level of carbon sequestration is many times the level we require for the world's degraded pasturelands and rangelands to reverse climate change over a thirty-year period. In other words, making the transition from industrial agriculture, monocropping, and intensive-confinement factory farming to agroecological, agroforestry, and silvopasture methods is absolutely essential if we are to not only mitigate but actually reverse global warming. This cannot be stressed enough.

Millions of farmers and ranchers across the world are already successfully utilizing carbon farming and ranching techniques. I'm personally quite familiar with a number of these practices because we use many of them on the two organic, transitioning-to-regenerative farms that I help manage, one in the semi-arid high desert region of north-central Mexico (6,300-foot elevation, with a four-month rainy reason and annual rainfall of twenty inches per year), and the other in the North Woods cold climate of northeastern Minnesota (seven months of winter and snow). In addition, our Regeneration International staff have visited hundreds of regenerative and transitioning-to-regenerative farms and ranches operated by our affiliates across the world. Carbon farming and ranching techniques are currently being carried out on hundreds of millions of acres across the world, in all climate zones and ecosystems, in different soils and terrain, in widely diverse regions with different rainfall and precipitation patterns, as well as different cultural and food traditions.

Carbon farming practices are not as unusual or complicated as they might seem, although they differ greatly from the degenerative, chemical-intensive, energy-intensive, monocrop farms and factory farms that corporate agribusiness and the mass media call "conventional agriculture." There are thousands of "how-to" articles, books, and videos in numerous languages describing carbon farming practices, sometimes categorized as organic agriculture, biodynamic farming, permaculture, agroforestry, conservation agriculture, holistic management, or just agroecology—all share the same common principles. There are colleges, universities, youth education projects, research institutions, government extension services, and farm and ranching organizations now actively engaged in spreading the theory and practice of carbon farming and soil health.

Even though we don't need to debate any longer whether organic or beyond organic carbon farming and ranching methods exist and successfully work, we do need to talk about how, in the time frame we have left before climate crisis turns into climate catastrophe (twenty to thirty years), we can scale up regenerative agriculture (along with reforestation and large-scale ecosystem restoration) enough to turn things around.

Carbon farming does indeed draw down much more carbon from the atmosphere, through enhanced photosynthesis, than industrial agriculture and factory-farm systems (the latter, in fact, are carbon and greenhouse gas emitters rather than carbon sinks). Carbon farming also generates healthier soil, crops, animals, fiber, and building materials (hemp, bamboo, et cetera), as well as nutrient-dense food that is free from chemicals, drugs, and GMOs. Carbon farming can and does improve the income and food production of small farmers and herders, urban gardeners, and even medium- and large-scale farm operations.

But the burning question remains: Can we popularize, create market demand for, finance, and scale up regenerative food and farming practices to such an extent that they actually begin to *reverse*, not just mitigate, global warming, while also providing other essential ecosystem, health, and economic benefits?

Reversing Global Warming

One proposal for how this can be done comes from Jim Laurie, a staff scientist for the public interest group Biodiversity for a Livable Climate. He calls it "Scenario 300: Reducing Atmospheric CO_2 to 300 ppm by 2061," and (in brief) he notes that it will require "good management on about half of available lands." As Laurie states:

> Scenario 300 proposes that restoration of degraded ecosystems of several types could capture enough carbon in soils to reduce the atmospheric CO_2 concentration from about 410 ppm in 2018 to 300 ppm before Halley's Comet returns in 2061. . . . Humans will have to make huge changes in the ways we interact with nature and with each other. Our actions must become less competitive as we work together to enable symbiosis in the recovery of massively degraded ecosystems. It will be the greatest challenge in human history and will require several decades to accomplish.[20]

In the final chapter of this book I will lay out a more detailed roadmap, using the United States as an example, for how our emerging Regeneration revolution can achieve net zero emissions in an advanced industrialized country by 2030.

As longtime organic farm leader Jack Kittredge points out in his popular science primer on reversing global warming, *Soil Carbon Restoration: Can Biology Do the Job?*, using just average carbon sequestration rates for carbon farms and ranches, not even taking into account exceptional or best practices, we can readily turn things around.[21]

As Kittredge explains, if we apply regenerative agriculture practices on the world's 8.3 billion acres of pastureland and rangeland (sequestering on average 2.6 tons of carbon or 9.5 tons of CO_2e per acre) and on the 3.8 billion acres of cropland (0.55 ton of carbon or 2 tons of CO_2e per acre), we could sequester 23.7 billion tons (gigatons) of carbon (87 billion tons of CO_2e) per year, more than twice as much per year as we are currently emitting. These average figures do not include the enhanced sequestration potential of high-fungal compost in croplands, agroforestry and silvopasture practices on croplands and pasturelands, or improved forest health (reforestation and afforestation) on the ten billion acres of the world's existing forests (including their native wetlands and peat bogs).

Of course, not every farmer and rancher is going to switch over immediately, or even in ten years, to regenerative practices. Millions of farmers are still profiting from chemical-intensive, industrial, factory-farm practices (especially in the United States and the European Union, where commodity farmers receive billions of dollars in government/taxpayer subsidies), while others appear just too stubborn to change. Given that we're not going to be able to change the practices of all farmers, perhaps it's more realistic to look at how we can regenerate a majority of the soils on this planet that are most degraded, which amounts to approximately 5 to 6.5 billion acres. These are the soils that governments, farmers, ranchers, and land managers would most readily agree should be regenerated. So, again, based on average (not the exceptional) sequestration rates, if we can regenerate these 5 to 6.5 billion degraded acres into healthy agroforests, pastures, rangelands, silvopastures, wetlands, and croplands (i.e., making them carbon sinks instead of carbon emitters) using already well-established carbon farming, carbon ranching, reforestation, and eco-restoration practices, we will have the ability to reverse global warming over the next thirty years (assuming we can reach zero fossil fuel emissions by 2050 or earlier).

In a series of ongoing and highly promising field tests, Dr. David Johnson of New Mexico State University has been able to sequester 4 to 7 tons of carbon (14.68 to 25.69 tons of CO_2e) per acre per year in US croplands, measured in just the first twelve inches of topsoil, utilizing biologically rich compost with a high level of mycorrhizal fungi.[22] We'll talk more about Johnson's work in the final chapter of this book.

Eric Toensmeier, in his book *The Carbon Farming Solution*, highlights even greater carbon sequestration potential in the world's tropical and semitropical areas, with intensive silvopasture and multiple canopy forest gardens sequestering between 6 and 14 tons of carbon (22 to 51 tons of CO_2e) per acre per year. Bamboo in areas with abundant rainfall, even raised as a monocrop, can sequester 4.84 to 13.8 tons of carbon (13.4 to 50 tons of CO_2e) per acre per year.[23]

Reforestation: A Necessary Complement to Carbon Farming

In addition to recarbonizing and regenerating agricultural lands, a major part of regenerating Earth and reversing climate change will be to preserve, restore, and expand the world's ten billion acres of forests and wetlands. This reforestation and afforestation will include planting up to a trillion tress in deforested areas (reforestation) as well as several hundred billion trees and perennials back into the world's four billion acres of cropland (agroforestry) and eight billion acres of pastureland or rangeland (silvopasture).

The current global tree population, which covers 30 percent of Earth's land area, is estimated to be three trillion trees, but fifteen billion trees are cut down every year. Since humans began farming ten thousand years ago, approximately half of the trees on Earth have been cut down and not replanted. Earth's forests and wetlands now sequester over seven hundred billion tons of carbon and currently draw down, even with massive deforestation and forest fires taken into account, an additional "net sink" of 1.2 gigatons of carbon."[24] The net sink or carbon sequestration power of today's forests equals approximately 12 percent of all current human emissions.

If "net deforestation" (more trees being cut down, clear-cut, or burned out than the number of trees growing to replace them) could be halted, especially in tropical areas where the trees grow faster and store the most carbon, and if forests worldwide could be managed to increase photosynthesis and biomass through massive reforestation and management (by thinning out overcrowded forest areas, transitioning from thousands of stressed trees per acre to hundreds of healthy trees per acre), the world's forests could net sequester four billion tons or more of atmospheric carbon a year, a full 40 percent of all current human emissions.[25] Along with renewable energy and carbon farming, if we stop deforestation and reforest the planet with a trillion species-appropriate trees, and then maintain these trees, we can reverse global warming.

The United Nations Environment Programme (UNEP) has now announced its Trillion Tree Campaign, with the goals of global reforestation and carbon sequestration. UNEP points out that there is enough deforested or empty space on Earth, in both rural and urban areas, to plant a trillion trees, of which six hundred billion mature trees can be expected to survive. And this trillion tree planting campaign does not include the additional one hundred-plus billion trees that could and should be planted on our twelve billion acres of croplands and grazing lands to take advantage of the tried-and-proven carbon-sequestering, livestock-friendly, fertility-enhancing techniques of agroforestry and silvopasture. More than 13.6 billion trees have already been planted as part of the Trillion Tree Campaign, which analyzes and projects not only where trees have been planted, but also the vast areas where forests could be restored. UNEP warns, however, that there are "170 billion trees in imminent risk of destruction," and they must be safeguarded for crucial carbon storage and biodiversity protection.[26]

UNEP's Trillion Tree Campaign was inspired in part by a recent study by Dr. Thomas Crowther and others that, integrating data from ground-based and satellite surveys, found that replanting the world's forests (totaling an additional 1.2 trillion trees) in the empty spaces in forests, deforested areas, and degraded and abandoned land across the planet would draw down a hundred billion tons of excess carbon from the atmosphere. According to Crowther:

> There's 400 gigatons now, in the 3 trillion trees, and if you were to scale that up by another trillion trees that's in the order of hundreds of gigatons captured from the atmosphere—at least 10 years of anthropogenic emissions completely wiped out. . . . [Trees are] our most powerful weapon in the fight against climate change.[27]

And Crowther's projections do not include the massive amount of carbon drawdown and sequestration we can achieve through agroforestry and silvopasture practices, planting trees, even if only a few trees per acre, in the world's often deforested four billion acres of cropland and eight billion acres of pastureland and rangeland.

In the United States, a massive program of reforestation, afforestation, and preservation of our forests, including planting billions of trees in our 254 million acres of degraded forests, as outlined in the Crowther study, could potentially sequester 1.5 billion tons of carbon by 2030 and even more thereafter. Incorporating agroforestry and silvopasture practices on our croplands and pasturelands could sequester many more billion tons of carbon. In other words, yes, we can achieve net zero emissions in the United States by 2030 and begin to provide support for the rest of the world to do the same. We will discuss US reforestation practices more in the final chapter of this book. But the time to begin this great reforestation is now.

No new technology or forest management techniques need to be invented. Techniques of passive or active forest restoration are practiced all over the world. Intact ancient or primary forests, amounting to 20 percent of global forested areas, are massive repositories of carbon and biodiversity. These forests need to be left basically intact, with subsidies and support for indigenous peoples and forest communities to sustainably protect and maintain them, as they have done for centuries. And, of course, there are more benefits beyond carbon sequestration; just as an example, ancient or old-growth forests provide habitat for two-thirds of all land-based animal and plant species on Earth.[28]

Passive restoration and selective thinning is all that is needed in many "second-growth" forest areas, stimulating trees, ecosystems, and wildlife to grow and thrive. However, in many forests more active restoration is necessary and, in fact, urgent, as today's massive wildfires and infestations by pests and diseases, such as pine beetles, ash borers, Dutch elm disease, and gypsy moth, attest. A global program of forest regeneration is necessary to speed up or help stimulate natural growth, increase biodiversity, and reduce the risks and intensity of wildfires. Thinning or controlled burns (not indiscriminate "slash and burn" methods) maximize natural tree growth, take out dead and diseased trees, promote biodiversity, and reduce excess ground-level biomass that fuels wildfires. Planned grazing of livestock in forests and forested areas, especially in forested areas adjacent to cities and towns, can substantially reduce fire risk while enhancing biodiversity and carbon sequestration at the same time.[29] Key species in this global reforestation effort will include fast-growing trees, like bamboo, in tropical and subtropical areas, and nitrogen-fixing coppice tree varieties, which reproduce and multiply after being pruned all the way to the ground, in temperate zones (alder, black locust, elderberry, chestnut) and tropical zones (acacia, moringa).[30] According to Eric Toensmeier's *Carbon Farming Solution*, multispecies woody polycultures in tropical and subtropical areas (cacao agroforests, woodlot polycultures, forest homegardens, banana plantings with other mixed fruits, and shade-grown coffee) can sequester more carbon (up to twelve tons per acre of carbon or forty-four tons of CO_2e per year) than any other type of agricultural system.[31]

Planting and nurturing native trees and bushes and transforming large sections of native forests into multi-strata, highly diversified food forests and forest gardens will need to be our goal. Understory crops can include vines, shrubs, bushes, and smaller trees (for example, coffee and other shade crops grown under highland tropical canopies, or hazelnuts and berries in the understory of temperate forests).[32] "Tree-range" livestock should also be integrated into forest environments. For millennia, indigenous people and small farmers have maintained food forests and forest gardens in large parts of the world, practicing permaculture and regenerative agriculture and forest management since long before the advent of so-called modern agriculture and industrial forestry.[33] Contemporary food forests are some of most productive, biodiverse, carbon-sequestering ecosystems in the world.

Forest Restoration and Biochar: Regenerative Synergy

Biochar, as noted earlier, is a special type of charcoal produced by pyrolizing (burning in a low-oxygen environment) wood and biomass. When added to the soil, biochar increases its fertility and water retention. Research indicates that biochar can increase food production by 15 percent or more, especially in the poor acidic soils that are routinely found in many poverty-stricken rural areas.[34]

In managed forests and landscapes around the world today, "excess biomass"—branches, underbrush, dead and diseased trees, and woody trimmings—is generally burned or tossed into landfills. Whether it burns or decomposes, it rises to the atmosphere as CO_2, contributing to global warming. Instead, as part of a global Regeneration strategy, all of that biomass must be converted to biochar and added to composts and soils.

As we've mentioned, biochar not only builds soil fertility but also sequesters carbon for long periods of time (some biochar in soils is thousands of years old). According to a number of biosequestration and forest experts, up to 30 percent of US emissions of CO_2 could be sequestered by pyrolyzing biochar from the excess biomass of the timber industry, forest slash, and crop residues (while taking care to avoid depleting croplands and forests of essential organic matter for the soil), as well as producing biochar from fast-growing biomass on idle farmland.[35]

The long-term potential of biochar for reducing human CO_2 emissions is enormous, especially if biochar stoves can start to replace traditional wood and biomass burning for cooking and heating in the Global South. Approximately three billion people still cook with wood and biomass (including dried cattle manure), using highly inefficient woodstoves and open fires. These wood fires are a major source of indoor and outdoor pollution as well as a major contributor to greenhouse gas pollution, emitting a billion tons of CO_2 annually, not to mention black soot. The World Health Organization estimates that 1.9 million people, mainly women and children, die each year from respiratory illnesses caused by smoke inhalation from wood cooking stoves and biomass burning. Specially designed biochar stoves could prevent the damage to both human health and the climate caused by burning wood and other biomass for cooking and heating.[36]

Today, the International Biochar Initiative is developing production standards and business support for responsible and regenerative biochar use.

The positive steps we need to undertake and scale up in forest restoration are rather straightforward and well understood. The Intergovernmental Panel on Climate Change (IPCC) has recognized for years that deforestation is responsible for a significant proportion (20 percent) of global-warming greenhouse gases. Unfortunately, there are major political and economic forces that are driving the degeneration of the world's remaining forests. The $320 billion global timber, paper, and packaging industry, along with agribusiness and factory farms, aided and abetted by corrupt politicians in countries like Brazil, Indonesia, Malaysia, the Congo, and other nations, compounded by mindless consumerism, are major driving forces in forest destruction and degeneration.

As we emphasized in the previous chapter, consumers must be educated in order for them to understand the serious climate impacts of their food and fiber choices and mobilize for political action and policy change. As examples, the mass destruction of tropical forests for GMO soybean cultivation and cattle raising in Latin America must be stopped. Consumers must be educated to boycott factory-farm meat and other GMO-derived foods, not only because these foods are bad for their health, but because these GMO crops, feeds, and factory farms are the number one factor destroying the Amazon and the natural grasslands of Argentina, Brazil, Paraguay, and other carbon-sequestering landscapes, which are the strategic carbon sinks of the planet. In the United States, more than 80 percent of our once-thriving prairie grasslands have been plowed up and planted with monocrops, mainly corn and soybeans for factory-farm animal feed and ethanol, along with monoculture wheat. Consumers must become aware of all the consequences of the consumer products they buy, whether food, fiber, timber, paper, body care, or cosmetics. Palm oil, coconuts, pineapples, coffee, cacao, soybeans, livestock, timber, and other agricultural products can continue to be produced and harvested in the tropics and other forest areas in a sustainable manner, but regenerative agroforestry and silvopasture practices, rather than clear-cutting, tree plantations, and monocrop production, must become the norm, along with a major reduction in the consumption of wasteful packaging and paper products, with 100 percent recycling becoming the norm rather than the exception.

Our forests, in all of the boreal, tropical, and temperate zones of the globe, are the veritable lungs of the planet, taking in vast amounts of atmospheric CO_2 and releasing life-giving oxygen. Trees are also the shade umbrellas and thermostats of the planet, shading understory plants, soil, forest animals, livestock, and urban environments and protecting people and all living organisms from excessive heat. They are also the air conditioners, air filters, and water pumps of the planet, absorbing pollution and filtering it through their leaves, while evaporating and transpiring radiant heat from the sun as moisture. This transpired moisture from trees is absolutely necessary for cloud formation and normal, predictable rainfall. The average plant on Earth uses less than 5 percent of the water absorbed by its roots for growth; the rest is released as moisture, a coolant for the planet. Besides regulating thermal heat and transpiring moisture into the air to help produce rain clouds, trees are our most important hydrologists and water suppliers: they regulate, store, and filter rainwater; absorb excess water; and store water in the soils and aquifers for plants, animals, and humans.

Trees also, of course, provide food, animal fodder, medicinal plants, biomass for cooking and heating, and building materials for over a billion people, as well as habitat for most of the world's biodiversity. According to a 2015 study, tropical forests alone provide habitat for two-thirds of all terrestrial plants and animals.[37]

Preserving our forests from industrial agriculture, overharvesting, mining, dam building, and other destructive activities is key, not only to our environment, but to our climate as well. But even beyond forest protection and conservation, we must *expand* our forests and wetlands and increase tree cover by implementing a vast global program of agroforestry and silvopasture, as called for in the United Nations' Trillion Tree Campaign.

Global Land Regeneration Priorities

The highest carbon sequestration rates in the world today are found in those lands with intact forests, wetlands, and peat bogs, as well as agricultural lands where agroforestry (especially multi-strata or multilayered agroforestry), perennial landscaping, and silvopasture practices have been implemented. Most of the degraded lands in the world that have the greatest capacity for increased photosynthesis, via forest and plant growth, lie in the Global South and the tropics, which is also where most of the world's poor people, small farmers, and forest dwellers live. As Eric Toensmeier points out, "The tropics have stronger carbon farming options than colder climates; many of the agroforestry techniques that have the highest sequestration rates . . . and the best perennial crops available today are also native to, or grown best in, the tropics."[38] This enhanced potential for carbon sequestration in the tropics and in the most degraded soils of the developing world means that the wealthier industrialized nations of the North must make it a top priority to help finance a massive scaling up of regenerative farming and grazing practices in the South. This targeted investment in regenerative food and farming will need to be accompanied by large-scale ecosystem restoration (which we will discuss shortly) as well as support for the preservation of tropical forests, wetlands, peat bogs, and marine ecosystems, many of which are located in indigenous communities.

Wetlands

Wetlands—forest, freshwater, riparian, boreal, alpine, and coastal—constitute, or at least once constituted, almost two billion acres on the planet. Wetland acreage is often categorized in climate studies as farmland or forest, and much of it has been degraded and drained by industrial agriculture and urban and coastal development. According to recent studies, 87 percent of global wetlands have been destroyed since the advent of modern agriculture, with 54 percent lost since 1900. Wetlands are able to capture and store up to four to six tons of carbon (14.7 to 22 tons of CO_2e) per acre per year, making them, along with peat bogs, extremely important as carbon sinks, more so than just about any other ecosystem. The reason for this is that wetlands are basically natural drainage areas where enormous amounts of eroded soil, leaves, brush, tree branches, and other vegetation accumulate and build up over time (think of river deltas, floodplains, riparian zones, lakeshores, coastal estuaries, or even just depressions in farm landscapes). That's a lot of sediment and nutrients—and carbon. Globally, wetlands are estimated to store 500 to 750 billion tons of carbon, almost as much carbon as is contained in the entire atmosphere.[39]

If one billion acres of wetlands could be properly protected, buffered, channeled, and allowed to naturally restore themselves, these restored wetlands could sequester six billion tons of carbon per year, while providing other important ecological services as well, such as filtering rainwater and providing wildlife and marine life habitat. As Jim Laurie of Biodiversity for a Livable Climate points out:

> The goal of restoring 1 billion acres of wetlands can be done if wetlands are incorporated into the plans for restoration of other systems. Wetlands can raise the water tables several feet in dry areas as we have seen in Zimbabwe and in Nevada, improving grazing opportunities there. Wetlands established in areas with higher rainfall can also improve resiliency of forests and farms. If 5% to 10% of these lands were managed as wetlands, the benefits would be enormous and [6 billion tons of carbon] would be captured, too. This percentage [of wetlands embedded in forests, farms, and other lands needing restoration] represents about 1 billion acres.[40]

Ecosystem Degradation, Desertification, and Restoration

Increasing rates of environmental degradation and desertification of once arable lands are now affecting the lives and livelihoods of 40 percent of the people on the planet. *Degradation* refers to all aspects of ecosystem services (soil carbon, soil fertility, food and forage production and quality, water availability, water quality, climate stability, air quality, biodiversity, et cetera), while *desertification* specifically refers to the degeneration of what was once arable, albeit semi-arid land into completely arid land, devoid of vegetation, water, and wildlife. Exact estimates vary, but globally there are approximately 1.5 to 3 billion acres (6 to 12 million square kilometers) of abandoned landscapes, deserts, and severely eroded areas that were once productive farmland, rangeland, and forested areas—sequestering carbon, producing food and fiber, regulating the water cycle, supporting livestock and humans—but now are nearly or completely dead. Meanwhile, steady degeneration is driving much of the world's agricultural and forested areas into desertification.

A perfect example of large-scale desertification is what was once called the "Fertile Crescent" of the Middle East. The Fertile Crescent is often described as the cradle of civilization, where agriculture (especially grain production and livestock grazing, based upon irrigation) and human civilization flourished for thousands of years. Today, though, vast areas of the Fertile Crescent in Iraq, Syria, Iran, Egypt, Israel, Palestine, Jordan, and Lebanon are desertified and no longer productive, mainly because of centuries of destructive agricultural and land use practices, such as continuous plowing, draining of wetlands, deforestation, overgrazing of livestock, and dam building. Of course, severe degradation and desertification are not limited to the Middle East; they are major problems in Asia, Oceania, Africa, Latin America, Europe, and North America as well.

Rather than dwelling upon the negative aspects of global landscape degeneration and desertification, it's important to understand that serious degradation and desertification can be mitigated, and in fact reversed, by local and even large-scale ecosystem restoration, the best recent examples of which have been undertaken in China, India, and Africa.

Perhaps the best example of large-scale ecosystem restoration (accomplished through reforestation, terracing, water catchment, and temporary removal of livestock) has occurred on the Loess Plateau of north-central China, at the headwaters of the Yellow River. In the Loess Plateau, an area the size of the Netherlands (640,000 square kilometers) of severely degraded mountain terrain, formerly productive forests, farmland, and grazing lands that had been reduced to a virtual moonscape were success-fully restored over several decades beginning in 1985. The project restored not just the regional ecology but also the livelihoods of five million Chinese farmers and their rural communities.[41]

One large-scale, desert-greening ecosystem restoration project currently being carried out is the "Great Green Wall" (GGW) in the Sahel region of Africa's Sahara Desert, whose planners are working to plant a swath of trees nine miles wide across the entire continent of Africa. By 2030 the Great Green Wall of billions of new trees will restore 250 million acres of desertified land and sequester 250 million tons of carbon. The reforesta-tion and soil-building practices being carried out along the Great Green Wall have been developed over decades on a smaller scale, based upon traditional and indigenous practices and modern permaculture innovation.

The GGW is an $8 billion reforestation and soil-building project sponsored by the African Union, the EU, the World Bank, and the United Nations Convention to Combat Desertification. According to a 2018 article by Interpress Service, the Great Green Wall "aims to restore Africa's degraded landscapes and transform millions of lives in one of the world's poorest regions."[42] The GGW will stretch across 4,700 miles and twenty countries, from Gambia in West Africa to Djibouti in East Africa. It will create thousands of new jobs and farming opportunities and improve food security for thirty million people who live in the Sahel, one of the driest areas in the entire world.

———————

The Loess Plateau and the Great Green Wall restoration projects are examples of the kinds of large-scale ecosystem regeneration that can and must be replicated on a global scale. Combined with regenerative farm-ing, grazing, and reforestation practices carried out on five to ten billion acres of farmland, rangeland, and forestland across the world, including a major scaling up of carbon farming practices such as agroforestry and silvopasture, large-scale ecosystem restoration will enable us, the global grassroots, to feed ourselves with healthy food, cool the planet, and stabi-lize the climate, before it's too late.

Politics and Public Policy

< regeneration driver three >

The Green New Deal we are proposing will be similar in scale to the mobilization efforts seen in World War II or the Marshall Plan. . . . Half measures will not work. . . . The time for slow and incremental efforts has long past.

ALEXANDRIA OCASIO-CORTEZ,
in a June 2018 e-mail to the *Huffington Post*

Social movements have an opportunity to join together as never before—not just to get behind the Green New Deal—but to form a broad-based, multi-racial, working class movement to build political power. Visionary leaders from these movements are already knitting together strategies for solidarity, education and action. . . . The Green New Deal just might be the fulcrum upon which the farm, food and climate movements can pivot our society towards the just transition we all urgently need and desire.

ERIC HOLT-GIMÉNEZ, "The Green New Deal"

The hour is late. We need a political revolution, in the United States and around the world, as well as a revolution in our food, farming, and land use practices. Our climate emergency certainly demands mass consciousness raising and decisive action in the marketplace on the part of consumers, along with a global scaling up of farmer innovation and regenerative land management, but we also need mass political mobilization and fundamental change in public policies. Fortunately, an important policy and legislative template for the political Regeneration that we need recently emerged in the United States: the Green New Deal (GND). The GND has galvanized significant media attention and public support, gaining the endorsement of more than a hundred Democrats in the US Congress, including prominent 2020 Democratic Party presidential candidates such as senators Bernie Sanders, Elizabeth Warren, Kamala Harris, and others.[1]

The GND was launched in 2018 by the youth-led Sunrise Movement and New York Congresswoman Alexandria Ocasio-Cortez. Sunrise's cofounder, Varshini Prakash, describes the GND as "an umbrella term for a set of policies and programs that will rapidly decarbonize our economy, get all of us off of fossil fuels and work to stop the climate crisis in the next 10 to 12 years."[2] Prakash told *Rolling Stone* magazine that the initiative has three pillars: 100 percent clean energy by 2030; investment in communities "on the frontlines of poverty and pollution"; and the guarantee of a quality job for "anyone ready to make this happen."[3]

The GND calls for a mass conversion to renewable energy and net zero emissions of greenhouse gases in the United States by 2030, policies in line with what most scientists now say are necessary steps to avoid catastrophic climate change. But what's unprecedented is that the GND also calls for system-wide economic Regeneration as well: jobs, health care, free public education, and income for all—policies extremely popular with the overwhelming majority of the body politic, including working-class and low-income groups, not just food and climate activists. The game-changing proposal combines, for the first time, a goal of 100 percent renewable energy, upgrading every residential and industrial building for state-of-the-art energy efficiency, a thorough decarbonization of industry and agriculture, and mass carbon drawdown and sequestration through regenerative practices.

The GND is still a work in progress, but it has made the American body politic sit up and take notice and put radical policy change on the agenda for the first time in a generation. Attacked as socialistic and impractical by the political establishment, its ambitious goals—addressing global warming and income inequality and transitioning to a green, fossil-fuel-free economy by 2030, while at the same time guaranteeing everyone who wants one a job with a living wage—have changed the political discourse of the nation.

The GND resolution introduced in the US House and Senate on February 7, 2019, includes the following specific passages relating to food, farming, and land use:

. . . working collaboratively with farmers and ranchers in the United States to eliminate pollution and greenhouse gas emissions from the agricultural sector as much as is technologically feasible, including—by supporting family farming . . . investing in sustainable farming and land use practices that increase soil health . . . building a more sustainable food system that ensures universal access to healthy food . . . removing greenhouse gases from the atmosphere and reducing pollution, including by restoring natural ecosystems through proven low-tech solutions that increase soil carbon storage, such as preservation and afforestation . . . restoring and protecting threatened, endangered, and fragile ecosystems through locally appropriate and science-based projects that enhance biodiversity and support climate resiliency . . . providing all people of the United States with access to clean water, clean air, healthy and affordable food, and nature.[4]

The brilliance of the GND is that it is bold and radical enough to actually address the overall climate crisis (net zero emissions by 2030), but populist and broad enough in scope to address the nation's economic and social crisis as well. In this way, the GND, unlike previous standalone climate proposals, has the potential to gain the support of the broad majority of middle-class, working-class, and lower-income communities, who are seeking good jobs, economic security, universal health care, affordable educational opportunities, and social justice, as well as a stable climate and clean environment.

Ideally, the GND will start to be implemented with a series of congressional bills after the 2020 elections in the United States, although lobbying for specific legislation that falls under the umbrella of the GND has already begun. Once the GND gains momentum and begins to be implemented in the United States, we can expect similar campaigns combining social justice, renewable energy, and regenerative food, farming, and land use practices to spread worldwide.[5] The bad news, however, is that we're running out of time. Unless we can quickly implement a program like the GND, as Ocasio-Cortez puts it, "similar in scale to the mobilization efforts seen in World War II or the Marshall Plan," the climate crisis will outpace our efforts.

Even though the current threat of the climate crisis is undeniable, there are powerful, entrenched politicians in both the Republican and Democratic parties in the United States who will not support a GND unless they are forced to by their constituents, or challenged by GND supporters in their primaries or general elections. As Naomi Klein points out, it will be necessary to build a mass-based movement outside the GND caucus in Congress in order to generate sufficient pressure for change.[6]

Underlining the urgency of bold and sweeping climate action, in November 2018 a global network of scientists, including the Intergovern-mental Panel on Climate Change (IPCC), warned the world once again that net emissions of greenhouse gases must be drastically reduced (by 45 percent at least) by 2030 if we are to avoid triggering deadly feedback mechanisms that will lead to catastrophic climate change.[7] The good news about the IPCC and global scientists sounding the alarm, however, is that the situation is now so dire that people, including politicians all over the world, are increasingly willing to listen and get behind bold new ideas and solutions such as the GND.

Net Zero Emissions by 2030 Requires Scaled-Up Regenerative Food and Farming

When GND proponents focus narrowly on reducing fossil fuel emissions and talk about achieving "zero fossil fuel emissions" by 2030, they invite criticism from skeptics who claim that this goal, however worthy, isn't feasi-ble. And in fact it probably isn't—if the plan to achieve zero emissions over the next decade is defined exclusively by the transition to renewable energy.

What *is* feasible? Achieving *net zero* emissions, through a combination of reducing emissions by transitioning to renewable energy and simulta-neously drawing down CO_2 from the atmosphere utilizing the power of regenerative agriculture and land use practices, including reforestation, wetlands restoration, and prairie and grasslands restoration.[8]

When discussing the goals of the GND, we must differentiate between "zero" emissions and "net zero" emissions. It may well be impossible to achieve zero emissions by 2030 (as opposed to "zero net emissions"), no matter how many resources we throw at developing clean energy tech-nologies, if that's all we do.

We need to do more. For the GND to accomplish its climate goals, it must spur two large-scale transitions: the transition away from fossil fuel use toward renewable energy, and the transition away from industrial agriculture (a huge polluter and greenhouse gas emitter in its own right) toward organic regenerative practices that draw down and sequester carbon.[9] The latter transition will have the added benefit of reducing emissions associated with industrial agriculture, which among other things include the enormous nitrous oxide and methane emissions from factory farms and synthetic fertilizer production and use.[10] Besides 100 percent renewable energy and system-wide energy retrofitting and conservation, as outlined in the GND, some of the specific regenerative food and farming policies that we need in the United States include:

1. Subsidies, tax incentives, minimum crop price guarantees (parity pricing), supply management (keeping supplies balanced with need so that prices are stable), antitrust enforcement, and marketing incentives for farmers, ranchers, and other land managers to improve soil health, increase soil carbon, adopt agroecological and regenerative practices (no-till farming, crop rotation, cover crops, planned rotational grazing, agroforestry, silvopasture) and other forms of ecosystem and watershed restoration, as well as ensuring the survival of family-scale farms and ranches.
2. Support for farmers to make the often difficult, multiyear transition to *certified* organic, holistic livestock management, and regenerative practices.
3. Changes in regulations making it easier for regenerative and transitioning-to-regenerative farmers (especially small producers) to process and sell their products—for example, meat and dairy products—not only locally (direct to consumer, retail, wholesale, institutional) but across state lines, regionally and nationally as well.
4. Public investment (and incentives for private investment) in regenerative food and farming infrastructure such as farmers markets, food hubs, processing facilities, storage facilities, marketing co-ops, and farmer training centers.
5. Loan programs and loan guarantees for land acquisition or leasing and infrastructure improvement for individual farms, ranches, food hubs, and urban agriculture projects, enabling young farmers, farmworkers,

and low-income individuals to buy land, livestock, and equipment and start their own farms.

6. Support for public procurement by government agencies and public institutions for organic and regenerative food and other products. Governments around the world spend trillions of dollars a year on public procurement for schools and other institutions, a significant percentage of which are food, fiber, building materials, and other agricultural products.

7. Support for school and campus organic gardens, local farm-to-table cafeteria programs, and development of school and university curriculums to stimulate student awareness and provide hands-on experience in regenerative food, farming, and land restoration practices, as well as food preparation, cooking, and nutrition education.

8. Subsidies for programs to provide jobs and job training in regenerative food and farming projects for youth and disadvantaged groups such as unemployed workers, women, and immigrants.

9. Taxes on carbon emitters and agrochemical companies to subsidize regenerative practices and projects.

10. Elimination of subsidies for degenerative energy, food, farming, and land use practices.

11. Lobbying cities, states, regions, and nations to sign on to and begin to implement the global "4 per 1000" Initiative to sequester excess atmospheric carbon and reverse global warming. Over three dozen nations, California, and a growing number of municipalities across the world have already signed on to the initiative.

12. Expansion and reform in the SNAP or food stamp system to promote the production and consumption of healthier, climate-friendly, organic and regenerative foods.

A number of these measures were introduced in September 2019 by Senator Cory Booker as part of his proposed Climate Stewardship Act.[11]

What Can the GND Do for Farmers and Consumers?

The US food and farm movement has long advocated for better policies, including better Farm Bill policies, that would support and incentivize farmers whose practices both draw down and sequester carbon and produce healthful, pesticide-free, nutrient-dense food. But given the urgency of the climate crisis, combined with the influence of industrial agribusiness and fossil fuel industry lobbyists over our political process, this approach—chipping away at Farm Bill policy every five years—provides little hope of achieving transformational, much less rapid change. We need radical changes, not just minor adjustments, in federal food and farm policies and annual appropriations after the 2020 elections, as well as in the next Farm Bill, which will come in 2023.

Elizabeth Henderson, a New York organic farmer and advocate for regenerative agriculture, says that the 2018 Farm Bill barely touches the structural and fairness issues that led to the current disaster for family-scale farms and the food security. Henderson recently wrote:

> The farm crisis of the 1980s that never really went away has resurfaced with a vengeance. In 2013, aggregate farm earnings were half of what they were in 2012. Farm income has continued to decline ever since. . . . Despite the shortage of farm workers, their wages remain below the poverty line. People of color and women are often trapped in the lowest-paying food system jobs and many are forced to survive on SNAP payments. The tariff game of #45 [Donald Trump] is only making things worse. The farm consolidation that has taken place has grave consequences for the environment and for climate change as well.[12]

Henderson argues that the GND's job and living wages guarantee must include jobs on farms: "If farms are guaranteed prices that cover their costs of production, farm earnings will be high enough to pay farm workers time-and-a-half for overtime over 40 hours a week, like workers in almost every other sector of the economy."[13]

The GND's guaranteed jobs and $15-per-hour minimum wage proposals would mean more income for farm workers and more money in consumers' pockets. Consumers would be able to buy more locally produced, nutrient-dense organic food. This in turn would generate more income for local farmers and food producers, who, under current economic conditions, increasingly are being forced into bankruptcy or having to sell out to big corporations with vast financial resources and economies-of-scale advantages.[14] The GND's call for universal health care also means that millions of farmers and food-chain workers wouldn't have to take on second jobs in order to provide health benefits for their families. This could help improve productivity.

And, of course, in addition to benefiting farmers, farm and food workers, and consumers, a new regenerative food economy will drastically reduce the emissions of the industrial food system, while at the same time building up healthy organic soils to provide a natural carbon sink.

Raising Global Consciousness: Grassroots and Grasstops

To get a regenerative GND in motion, in the United States and beyond, will require a monumental effort, including an unprecedented consciousness-raising and coalition-building campaign. Regenerators everywhere must get our positive message out to a mass audience so that we are constantly recruiting new advocates and volunteers. But even beyond the rank and file, we must also get our Regeneration message out to *grassroots influentials*, or "grasstops." Grasstops are those individual community members, including farmers, chefs, scientists, teachers, businesspeople, women, church leaders, labor activists, health practitioners, artists, students, and progressive-minded politicians, whose voices have the extra power to reach and activate a mass audience. Mobilizing grasstops, along with the grassroots rank and file, will provide us with a qualitatively stronger voice in our communities, allowing us to impact not only the marketplace but politicians and public policy as well. With massive grassroots support, amplified by grasstops' influence, we must put regenerative food, farming, and land use on the table for political discussion in every one of the cities and towns in the United States and across the world, building maximum support by linking these issues to the strategic "hot button" economic, social, and environmental issues of local constituencies and communities (jobs, income, health, environment, justice, et cetera).

Our global consciousness-raising campaign will require us to approach and build local, state, and national coalitions with a multitude of civil society and political organizations who have their own preexisting issues and agendas and may not yet fully understand the Regeneration paradigm shift. This is why a far-reaching, holistic economic/social justice/climate frame and template such as the GND is so important. We must carefully craft the content and style of our messages and educational materials and carefully select our public messengers, including influential grasstops. We must also broaden and amplify our message (and our power) by uniting radical, liberal, and conservative voices wherever possible, both rural and urban, though without sacrificing our basic principles and goals. Many individuals and groups concerned about jobs, income, health, justice, sustainability, peace, government corruption, and other issues are starting to grasp the deadly seriousness of the climate crisis, making it easier to get them to connect the dots between the issues that they already understand and the regenerative food, farming, and ecosystem restoration solutions that we need. Once we build new alliances with broader segments of the body politic, we can then more effectively lobby, influence, and elect (or vote out of office) politicians to generate public policy change.

Regenerative Practices Are Good for Everyone

Regenerative practices and policies are good for everyone, everywhere. Embodied in public policy, legislation, economic planning, foreign relations, and international treaties, regenerative practices can socially and economically revitalize rural and urban communities alike, eliminating much of the poverty, disease, unemployment, hopelessness, and desperation that are now driving forced migration, organized crime, drug addiction, terrorism, violence, and war. There is no downside for candidates for office and elected public officials to speak out for healthier soil, food, and farm animals, for green jobs, and for all the other benefits of regenerative farming and land use and an overall GND (except perhaps losing donations from vested special interests upholding the degenerate status quo). Most politicians, left, right, and center, will understand and support the benefits of Regeneration—especially once they are educated and lobbied by their constituents, constituents who stand to benefit directly from the policies we are talking about. Even highly centralized or authoritarian governments, such as Russia, China, Iran, and India, understand the existential imperative of averting climate catastrophe, especially if doing so allows them to simultaneously strengthen their economies, become more self-sufficient in terms of food production, and placate their ever more restive populations at the same time.

To put our innovative solutions on the table for discussion, all the way from city council and school board meetings to congressional races and the presidential debates, we must carry out a wide-ranging program of public education, citizen lobbying, and coalition building, starting with anyone and everyone willing to listen. Organizationally, we will need to build core advocacy groups, networks, and lobbying alliances in every community and region, utilizing organizational vehicles such as the GND, wherever possible. These Regeneration networks are already starting to materialize across North America and other parts of the world, including networks in California, Iowa, the Mid-Atlantic, Minnesota, Missouri, Nebraska, New England, Washington, Canada, Belize, Chile, Guatemala, Mexico, Africa, Australia, and the Philippines. One of the coalitions that I have been working on in the United States is Farmers and Ranchers for a GND. Another is called Green Consumers for a GND.

As we gain converts and create a common understanding, representatives of our Regeneration and GND coalitions—farmers, workers, unions, church congregations, students, environmentalists, senior citizens, environmental and social justice organizations, progressive businesses—can unite our forces and approach elected public officials and candidates running for office at every level: town board, city council, school board, county government, state government, and federal politicians. Our rule of thumb should be to first approach those with an open mind, and not waste our time on bought-and-sold politicians who are completely beholden to fossil fuel corporations and corporate agribusiness, or who deny that climate change is a problem.

When we lobby politicians in their home districts or state and national offices, we must strive to build a citizens' lobbying team that includes a wide variety of stakeholders. One of the most successful tactics we can use is to bring politicians and delegations out to the countryside to see regenerative farms, ranches, and ecosystem restoration projects firsthand, or to bring them to farm-to-table restaurants, school and urban agriculture projects, or farmers markets. During these on-site visits, politicians and their staff advisors can speak with farmers and landscape managers about what they are doing, sample regenerative food, and discuss the specific legislation and public policies that we need to support Regeneration.

But there are major obstacles standing in the way of our consciousness-raising and coalition-building campaign. Corporate special interests and their bought-and-paid-for indentured politicians are opposed to any system change, such as the GND, that would reduce their inordinate profits and power. In addition, most people, including politicians, have not yet heard about the world-changing power and promise of regenerative practices. The majority of ordinary people, driven by necessity, tend to stay focused on their own everyday problems and concerns. Understandably, they may feel that the issues that most directly impact them, like poverty, jobs, hunger, unemployment, social justice, education, deteriorating public health, or peace, are more important than climate stability. The majority of the people in the world are struggling so hard to survive that they don't have much time to think about climate change, much less go to a meeting or join up with the Regeneration movement. Today, approximately 80 percent of people, even in a wealthy country like the United States, are living from paycheck to paycheck, while several billion people in the Global South—farmers, workers, and consumers—are living in poverty or near-poverty. This is why we must always connect the dots between the regenerative solutions we are talking about and the practical, pressing concerns of everyday life. This is why bold, multi-issue, radical populist approaches such as the GND are all-important.

———

The Regeneration revolution will not unfold in a day, or even a year. It may well take a decade or more to get a critical mass of people in a critical mass of communities to connect the dots between their individual concerns and get them to lobby and push for the collective, indeed global, solutions we are offering. The global Regeneration movement we need, agronomic and political, will likely grow and spread slowly over the next few years, gathering strength and momentum, but then, as the climate and economic crises gets worse, our movement has the potential to rapidly expand. Once it becomes clear that a global drawdown strategy and a just, green economy is our only hope for survival, our ranks will grow by the millions. But the time to begin this long campaign of public education, network development, coalition building, policy development, and grassroots lobbying is now.

What is crucial is not that everyone, everywhere, immediately agrees 100 percent on all of details of the Regeneration revolution or the GND—that is just not practical—but rather that we build upon the proven examples of renewable energy and organic, agroecological, and regenerative practices and the validity of our basic paradigm, broadening our movement to embrace the shared concerns of the majority in each community, region, nation, and continent. Through a diversity of messages, frames, and campaigns, through connecting the dots between all the burning issues, we will find the strength, numbers, courage, and compassion to build what will eventually become the largest grassroots coalition in history—a united front and a GND to safeguard our common home, our survival, and the survival of future generations.

The Oncoming Green Wave in North America

Our political economy and culture, in most nations and communities, have in recent decades fallen under the control of powerful special interests. Corrupt and indentured politicians, aided and abetted by large corporations, have routinely hoodwinked and divided the body politic, mismanaged public resources, and propped up a deadly, climate-destructive energy, agricultural, and economic system. But as a growing sector of the United States and global body politic, especially the younger generation, understand, we must find a solution to global warming, climate change, economic justice, and other pressing issues. Although the overwhelming majority of the 195 nations of the world and the million communities, cities, and towns that make up these nations have not yet taken decisive action to scale up regenerative food, farming, and land restoration to complement their growing investments in renewable energy, surveys and political trends indicate that a wave of major progressive and green-minded change is coming, as exemplified by recent developments in the United States and Mexico.

The 2018 elections in Mexico and the United States provide positive evidence that the political pendulum is shifting away from business as usual, and that climate denial, climate inaction, and government by and for the rich and the multinational corporations may be coming to an end. In Mexico, Andrés Manuel Lopéz Obrador and his Movement for National Regeneration (MORENA) captured control of all three branches of government, promising to put an end to seven decades of corrupt elite rule and to deal with the climate crisis, the agricultural crisis, the drug war, and the economic crisis of the country—putting the interests of small farmers, workers, migrants, indigenous communities, youth, and low-income constituents and overall public health, food sovereignty, and environmental concerns ahead of the interests of multinational capital. Obrador's ministers, among other commitments, have reiterated their support for the "4 for 1000" Initiative, which Mexico has signed, and have called for a revitalization of traditional agriculture and an extensive program of reforestation and ecosystem restoration. Mexico's ballot box revolution and the program of MORENA will likely inspire and embolden similar progressive green waves throughout Latin America and the rest of the Global South.

In the November 2018 elections in the United States, the scandal-plagued Trump administration and the Republican Party lost control of the House of Representatives and a number of statehouses, after pushing a legislative agenda that included climate change denial, rollbacks of environmental and climate safeguards (which had regulated carbon emissions from coal power plants and increased automobile fuel efficiency standards, among other reforms), sharp cutbacks in health and social services, a crackdown on immigrants, expansion of our already swollen military budget, and multitrillion-dollar tax cuts for the rich. One important trend among the US body politic that became evident in 2018 is that over two-thirds of young people and nonwhite ethnic communities (Latinos, Asians, Muslims, and African Americans) voted against the Republican agenda, electing a new progressive and liberal majority in the House, a growing number of whom, led by Alexandria Ocasio-Cortez of New York and other Latinas, African Americans, Muslims, and Native Americans, have endorsed the notion of a GND. This new wave of insurgent voters and politicians, arising largely from millennials and ethnic communities, represent the majority of the body politic that will determine the outcome of elections and public policy in the crucial coming decades. This progressive and green-minded majority, supported by suburban women and college-educated workers, now have gained the upper hand in powerful states like California (which signed on to the "4 for 1000" Initiative in 2018) and cities like Portland, Oregon (which passed a clean energy ballot initiative in November 2018 to force billion-dollar corporations to pay for renewable energy and regenerative programs).[15]

Although the majority of congressional Democrats (and all Republicans) have yet to come out in favor of a GND of economic justice, 100 percent renewable energy, and carbon sequestration, a clear majority of registered Democrat voters, and even Republican voters, are now expressing concern over the climate crisis. A growing number of congresspeople (in both parties), along with numerous state and local officials, are now speaking out in favor of soil conservation, rotational grazing, agroecology, taxes on fossil fuel emissions, and ecosystem restoration. In 2016, democratic socialist Bernie Sanders, a supporter of GND policies, nearly gained the Democratic nomination for president, with polls predicting he could have defeated Donald Trump in the general election. With a Democratic president in the White House and a progressive majority in both the House and the Senate, the US GND could become a reality, setting a powerful precedent for the whole world. Despite setbacks, similar green and left populist waves are gathering momentum in the UK, Greece, Spain, and France, while climate-conscious social democrats and green-minded progressives continue to wield significant influence in Germany, France, and the Scandinavian countries.

Nevertheless, most political progressives, and liberals, while stronger on social, environmental, and economic justice issues than their politically conservative counterparts, still appear to be ill-informed on matters of food, farming, nutrition, natural health, and Regeneration, even as they express concern over the climate crisis, deteriorating public health, economic injustice, and environmental degradation. In fact, in the United States, in certain cases conservatives and libertarians, especially those representing rural districts, appear to be more enlightened on certain food and farming issues than their supposedly green-minded Democratic Party counterparts. The PRIME Act, for example—a federal legislative proposal to help small farmers sell their locally produced, state-inspected meat and animal products processed in smaller, more affordable local slaughterhouses—was reintroduced in 2019 and attracted support from both Republicans and Democrats. Industrial hemp, finally legalized in the 2018 Farm Bill, was long championed by Republican members of the House and Senate, including some arch-conservatives. Meanwhile, many conservative ranchers and farmers, as well as progressives, have begun adopting regenerative grazing and soil conservation practices.

Reading over the documents and policy platforms of various political parties and organizations in the United States, like Our Revolution, Justice Democrats, Brand New Congress, Indivisible, Democratic Socialists of America, the Progressive Caucus, and the Democratic and Republican parties, there is, except for the notable exception of the GND, currently little or no mention of regenerative practices, soil health, and ecosystem restoration, and only a minor discussion of food and farming issues in general. In fact, in the United States and Europe, even major environmental and climate groups such as Greenpeace and 350.org have not yet clearly spoken out on Regeneration issues, although organic farm, consumer, and industry organizations have begun to embrace Regeneration as the next necessary stage of organic food and farming. The same goes for most political parties in Europe (with the exception of the Green Party) and other nations. A major challenge for the Regeneration movement will be to engage with and win over a critical mass of politicians during the next few years, with a special emphasis on the new wave of green and progressive politicians who appear poised to take and maintain power in North America and other nations. But again, there is no downside for politicians of all stripes, from libertarians to leftists, to speak out in favor of regenerative practices. Our issues and our solutions are universal, with appeal for rural and urban consumers, farmers, and minority and frontline communities as well.

Redefining National and International Security

Regeneration necessarily implies, in geopolitical terms, redefining national and international security. To avert catastrophe, the body politic in every nation will need to move beyond the divisive themes of nationalism, empire building, and ruthless competition and conflict over natural resources to a cooperative global perspective that is focused on our current climate emergency and the mortal threat to our common survival.

Without global cooperation and diplomacy and respect for human rights and international law, there will be no chance for Regeneration to take hold. Geopolitical power plays must come to an end. Either we have a multitrillion-dollar military-industrial complex, endless war and confrontation, and a degenerative globalized factory-farm food system, or we have a livable planet with stable and predictable weather, a healthy environment, nutritious food, and economic livelihoods for all. But we cannot have both. Diplomacy must replace saber-rattling and military action on every continent, putting an end to the raging wars in Syria, Afghanistan, Iraq, Yemen, Israel-Palestine, and elsewhere with a dismantling or refocusing of NATO, a closure of US bases overseas, and an end to the superpower conflicts between Russia, China, Europe, and the United States.

The biggest threat to our common security, to our collective survival, as even the Pentagon has admitted, is neither war nor drug trafficking nor terrorism but, rather, runaway global warming. Paraphrasing a 2014 report from the Pentagon:

> [The] U.S. Department of Defense is "already beginning to see" some of the impacts of sea level rise, changing precipitation patterns, rising global temperatures, and increased extreme weather—four key symptoms of global warming. These symptoms have the potential to "intensify the challenges of global instability, hunger, poverty, and conflict" and will likely lead to "food and water shortages, pandemic disease, disputes over refugees and resources, and destruction by natural disasters in regions across the globe."[16]

A major challenge of the Regeneration movement will be to change the global discussion on foreign policy in each of our countries from its current obsession with exports, gross national product (GNP) growth, and the so-called wars on terrorism and drugs to address the biggest threat to our national and collective survival: global warming. Only by changing the conversation and refocusing public consciousness will we be able to revitalize the international peace movement, deal with the underlying causes of rural poverty and forced migration, and reduce sectarian strife. Only by focusing on regenerative solutions to the climate crisis will we be able to bring about food security and a decent level of economic prosperity in rural communities, where half of the world's population and most of the poor live, and where much of the "heavy lifting" of eco-restoration and soil carbon sequestration will need to take place.

One important theme to emphasize globally is that North/South cooperation and solidarity, and especially support for small farmers and forest communities in the Global South, is the key component in sequestering enough soil carbon to restabilize the climate. The billions of acres of degraded soils of the Global South (Asia, Africa, and Latin America) and the still biodiverse tropical and semitropical forest and grassland areas of the developing world hold the greatest potential to sequester soil carbon on Earth. No matter how good a job we do in the United States and Europe in regenerating our soils, sequestering carbon, protecting our forests, and converting to renewable energy, it will not be enough to solve the climate crisis. By scaling up agroforestry, silvopasture, biodiversity, and perennial crops in the Global South, and protecting and restoring tropical forests, grasslands, wetlands, peat bogs, and marine ecosystems in these areas, we will be able to draw down and sequester most of the several hundred billion tons of excess atmospheric carbon that we need to mitigate and reverse global warming.

Similarly, in the massive boreal forests and rapidly melting, methane-rich permafrost areas of Russia, Scandinavia, Canada, and Alaska, we will need to turn away from geopolitical competition, trade rivalries, and natural resource extraction to cooperative ecosystem restoration (including reforestation, afforestation, regenerative grazing by caribou, reindeer, elk, deer, and bison, and the use of biochar soil amendments) or else face mounting, potentially catastrophic CO_2 and methane releases from forest fires, melting permafrost, and diseased, dying, or clear-cut forests. Twenty-four percent of the northern hemisphere is currently covered with permafrost, precariously sequestering 1.8 trillion tons of carbon, more than twice as much as is currently contained in Earth's atmosphere. If the countries of the far North continue to fight among themselves instead of cooperating to save the boreal forests, reintroduce herds of grazing animals, utilize biochar, and reverse permafrost melt, it may become impossible to stop a gigantic methane release that could make our entire planet uninhabitable.

Global North cooperation with and support for small farmers and herders in the Global South and tropical agriculture and ecosystem restoration will require a major decrease in international tensions and an increase in global solidarity and cooperation to get the job done. Thousands of nongovernmental organizations (NGOs), foundations, governments, and international funding institutions are already engaged in building and expanding local, national, and cross-border food, farming, conservation, and alternative energy projects, providing living examples of international solidarity and Regeneration. But current levels of North/North, North/South, and South/South solidarity and cooperation will have to increase significantly over the next decade and beyond if we are to survive our climate emergency and thrive.

To safeguard our survival, we must put an end to the wars over fossil fuels and strategic resources, including water and precious minerals, and begin to address and repair the degraded soils, watersheds, grazing lands, marine ecosystems, tundra, and forests that underlie so much of the poverty, hunger, and desperation driving mass migration, terrorism, and war. In short, we're going to have to build a global coalition of Regeneration, peace, social justice, and economic justice advocates powerful enough to convert the multitrillion-dollar war machine and armies of the world into a mighty force for peace, disaster relief, Earth repair, biodiversity restoration, carbon sequestration, and climate restabilization. And, of course, at the same time we will have to move away from energy- and chemical-intensive industrial agriculture and factory farms and destructive land use by timber and mining corporations.

We need a new twenty-first-century redefinition of national and global security, one that focuses on the mortal threat to our species, runaway global warming, and catastrophic climate change. The only *just war* is the war against climate change. As *Solutions* journal put it in 2012: "National security is no longer just the practice of containing foreign threats. It also requires that we strengthen the fabric of society—our communities, infrastructure, economies, health, and social systems."[17] In terms of economics, global security necessarily implies that we turn away from "profit at any cost" business practices that serve only to increase the already exorbitant riches of the 1 percent to a new marketplace dynamic that maximizes "natural capital": soil, biodiversity, and environmental health; consumer health and well-being; and the survival and prosperity of the world's three billion small farmers, pastoralists, indigenous communities, fishing communities, and rural villagers. A new international majority, a survival-and-revival coalition comprised of billions of consumers, farmers, scientists, businesspeople, religious communities, and activists, must deliver the urgent and redemptive message that a livable climate and environment trumps so-called "free trade" and the maximizing of short-term corporate profits.

Fortunately, even bottom-line businesses and investors are starting to see the value of energy conservation and renewable solar and wind over fossil fuels. As Amory Lovins of the Rocky Mountain Institute and other energy analysts point out, conservation and renewable energy will replace fossil fuels by 2050, not only because this will be better for the climate and the environment, but because it will prove to be more profitable in the long term to corporations and businesses, even the fossil fuel giants.[18] A similar enthusiasm for regenerative practices will soon emerge as businesses and investors realize that organic and regenerative food and farming are excellent investments, and that factory farms and industrial monocultures, on the other hand, are profoundly dangerous in terms of fueling climate change. Once this understanding takes hold, we will see investors and corporations start to divest from industrial agriculture, factory farms, and nutrition-deficient, processed food and reinvest in organic and regenerative food and farming.

Regeneration advocates certainly have no problem with honest businesses, farmers, or land managers making an honest profit, nor with fair trade, respect for labor, and equitable markets that generate reasonable profits. Rather, our Regeneration rebellion is directed against unethical business and marketplace practices that treat people, health, the environment, the soil, and essential biological and climate-regulating systems as mere "externalities" or secondary concerns.

Some pessimists argue that the Global South (Asia, Africa, and Latin America), where most of the world's population lives, is too preoccupied with moving beyond poverty and creating jobs to put a priority on reducing emissions, ecosystem restoration, and reversing global warming through enhanced photosynthesis and natural sequestration. But the extraordinary thing about decarbonizing food and farming, restoring grasslands, and reversing deforestation—moving several hundred billion tons of carbon back from the atmosphere into our soils, plants, and forests—is that this Regeneration process will not only reverse global warming and restabilize the climate but also regenerate local, regional, and national economies. Global Regeneration will stimulate hundreds of millions of rural (and urban) jobs, while qualitatively increasing soil fertility, water retention, farm yields, air and water quality, and food quality. China is already becoming a leader in ecosystem restoration as well as renewable energy. India and African nations are pushing forward on reforestation, while Russia, Western Europe, and the regional governments of the Himalayan region are moving to make organic farming the norm.

Our global political, consumer, farmer, and investment Regeneration revolution must be organized and mobilized simultaneously. Neither component of this global transformation can succeed without the other. We must vastly increase our pressure on governments and corporations to change public policies and marketplace practices, but we must also identify, support, and elect new politicians to office who support our new Regeneration revolution and a GND. We need every nation, province, city, and village in the world to endorse and implement a program of Earth repair and healthy soils. The French "4 per 1000" Initiative put forth at the 2015 Paris climate summit is a good place to start.

The "4 per 1000" Initiative: A Global Plan to Reverse Global Warming

Beyond the standard gloom-and-doom talk on climate, one of the most important and promising developments in the world today is the "4 per 1000" (or 4/1000) Initiative put forth by the French government for the first time on the global stage at the UN climate summit in Paris on December 1, 2015. The 4/1000 Initiative calls on the nations of the world not only to move as quickly as possible to 100 percent renewable energy, as pledged at Paris, but to also start drawing down and sequestering into the world's soils as much carbon and greenhouse gases from the atmosphere as nations are currently emitting, and to continue this drawdown for the next twenty-five years (from 2020 to 2045), utilizing agroecological and regenerative food, farming, and land use practices. An average drawdown or increase of only 0.4 percent (4 parts per 1000, and hence the name of the initiative) of carbon in the world's agricultural soils and forests each year over twenty-five years, in combination with 100 percent (or near 100 percent) renewable energy, will basically enable us to restabilize the climate and begin to reverse global warming. Successfully carrying out the 4/1000 Initiative can move up to 250 billion tons of atmospheric carbon into our living soils, reducing CO_2 levels in the atmosphere from our current dangerous level of 415 ppm to 280 to 350 ppm.[19]

The 4/1000 Initiative has now been endorsed by almost three dozen nations, including Denmark, Finland, France, Germany, Sweden, the UK, Australia, New Zealand, Canada, Brazil, Costa Rica, Mexico, Uruguay, and Morocco, as well as the UN Food and Agriculture Organization (FAO), the state of California, and several hundred civil society organizations. Proponents of 4/1000 expect most nations, regions, and cities will sign on to the initiative over the next few years, enabling them to meet their NDC (nationally determined contributions) obligations under the Paris Climate Agreement. The United Nations Framework Convention on Climate Change (UNFCCC) recognizes the "4 per 1000" Initiative as part of the Lima-to-Paris Action Agreement.

In terms of carbon sequestration on pasturelands and rangelands, Dr. Richard Teague of Texas A&M University believes that current regenerative grazing practices in the United States and elsewhere show that the 4/1000 goal of a 0.4 percent yearly increase in soil carbon is quite practical:

> Data from leading conservation ranchers in North America indicates that with appropriate grazing management the goal of the COP21 Climate Summit in Paris to increase soil carbon on grazed agricultural land by 0.4% per year can be exceeded by a factor of 2 or 3. With appropriate grazing management, ruminant livestock consuming only grazed rangeland and forages can increase carbon sequestered in the soil to more than offset their greenhouse gas emissions. This would result in a GHG-negative footprint, while at the same time supporting and improving other essential ecosystem services for local populations. Affected ecosystem services include water infiltration, nutrient cycling, soil formation, reduction of soil erosion, carbon sequestration, biodiversity, and wildlife habitat.[20]

Mounting public concern is the major reason why 195 nations of the world signed the Paris Climate Agreement in 2015, pledging to eliminate net fossil fuels emissions by 2050, with the goal of keeping the average global temperature from rising more than 1.5°C (2.7°F). This is also the reason why several dozen nations have already signed on to the "4 per 1000" Initiative as well.

But, of course, mounting public concern over global warming has not yet translated into concerted political action and policy change. Institutional inertia, business as usual, government corruption, and the power of entrenched special interests (the fossil fuel and agribusiness sectors in particular) have stalled strategic reforms. Although our Earth crisis demands a rapid and comprehensive global solution, most national governments, including that of the United States, are still pushing energy, conservation, farm, food, and land use reforms that amount to "too little, too late." While Big Ag, Big Biotech, Big Pharma, Big Oil, Big Timber, Big Finance, Madison Avenue, and the military-industrial complex lobby and fund degenerative public policy to maintain business as usual, the delicate climate-stabilizing balance between carbon levels in the soils, atmosphere, and oceans of our fragile Earth unravels. National pledges or NDCs signed by the nations of the world in Paris in 2015 to reduce greenhouse gas emissions to net zero by 2050 or earlier and the 4/1000 pledges to sequester billions of tons of atmospheric carbon every year have not yet, for the most part, been implemented. In addition, industrialized nations' pledges to donate $100 billion per year to a Green Fund to help developing nations achieve zero emissions have not materialized, with a pro–fossil fuel Trump administration in the United States actually threatening to pull out of the Paris Climate Agreement altogether in 2020.

In 2019, CO_2 concentrations in the atmosphere reached a record-breaking 415 ppm, with a corresponding increase in average global temperatures, driving alarming feedback trends, including the accelerated melting of the polar ice caps, decreased summer ice in the Arctic, sea level rise, biodiversity loss, severe damage to coral reefs, violent weather (droughts, storms, floods), forest fires, soil erosion, crop deterioration and failures, water shortages, and spreading disease and pestilence, affecting plants, animals, and humans. In the summers of 2018 and 2019, a massive heat wave engulfed much of the planet, with scientists attributing this phenomenon to disruptions in the jet stream over the Global North, triggered by rapid warming in Arctic regions.[21] This massive heat wave in turn ignited massive, deadly forest fires in California at the end of 2018 and again in 2019.

Can We Afford a Green New Deal?

The GND, as the youth-led Sunrise Movement reminds us, is the moral equivalent of war—a just war that we cannot afford to lose. And yet a major criticism of the US Green New Deal and its stated goal of reaching net zero emissions by 2030, by Republicans and neoliberal Democrats, is that it will cost too much. Of course, haggling over the price tag to safeguard our literal survival, and the survival of our children and future generations, is a form of lunacy and moral degeneration. As the Levy Institute puts it: "The costs of extinction of the human species—from the point of view of humans, at least—is beyond measure. Even if we calculate the costs of the GND as $93 trillion (as one hysterical estimate puts it) over the next decade, that is puny in comparison with the discounted cost of total destruction of human life on planet Earth."[22]

The only real choice we have left is to spend whatever it takes to avoid climate catastrophe and global societal meltdown, just as our parents and grandparents in World War II spent whatever was necessary to defeat the Nazis. The critics' cry that the GND "will cost too much" is nothing more than a delusional last gasp to maintain business as usual, to sustain the profits, power, and privileges of the 1 percent.

The United States spent $4 trillion in 1941–45 (in today's dollars, adjusting for inflation) to fight World War II, with expenditures rising to 40 percent of the gross national product in 1945.[23] In comparison to the 1945 GNP of $2.3 trillion (in today's dollars, adjusting for inflation), the current US GNP is $19 trillion, and that number is projected to rise to $33 trillion by 2030.[24] Forty percent of those two figures—matching the percentage the United States spent in a single year to fight World War II—is $7.6 trillion and $13.2 trillion, respectively.

According to Fox News and a chorus of conservative pundits and Republican political leaders, the GND will supposedly cost US taxpayers, businesses, and the government $93 trillion over the next decade, bankrupting the entire economy.[25] Analysts, however, have pointed out that this $93 trillion figure is wildly exaggerated because it includes over $70 trillion in health care and other social program costs (job creation, job retraining, raising the minimum wage, education, social services) that will likely be spent over the next decade anyway, as if these would be *extra* costs of the GND.[26]

For example, US health care costs (for consumers, taxpayers, businesses, and the US government) amounted to $3.65 trillion in 2018[27] and are expected to rise steadily (5 to 8 percent per year) over the next decade to reach almost $15 trillion annually by 2030, if our current for-profit, Big Pharma business-as-usual model is left in place.[28] Bernie Sanders's "Medicare for All" plan, in comparison, with a focus on improving overall public health, cutting out corporate profits, controlling drug costs, and reducing administrative overhead, as outlined in the GND, is estimated by Sanders to cost $40 trillion over the next decade, trillions of dollars less than what taxpayers, businesses, and the government would be spending under a business-as-usual scenario.[29] In other words, not only can we afford a GND, but we will likely bankrupt the entire economy if we *don't* implement a GND.

In any case, the actual costs to rapidly move to renewable energy, radical energy conservation, and regenerative food, farming, and land use practices, so as to achieve net zero emissions by 2030 in the United States, will likely turn out to be in the range of $2 trillion in additional public funds a year, not $9 trillion a year, as critics of the GND claim. Bernie Sanders's program for a GND adds up to $16.5 trillion over ten years.[30] This $2 trillion a year in loans, bonds, and subsidies will enable us to reduce current fossil fuel emissions by 50 percent by transitioning to a green renewable economy, while simultaneously drawing down and sequestering the equivalent of the remaining 50 percent in our soils, forests, and plants by scaling up regenerative food, farming, and land use practices, as outlined in the final chapter of this book. (This $2 trillion a year in additional public funds does not include the trillions of dollars that the private sector will likely be investing in a new green economy over the next decade.)

But how can the federal government, assuming a change in political power after the 2020 elections and beyond, come up with an additional $2 trillion a year to fund the GND? Besides levying a tax on fossil fuel polluters to force them to move to renewables, transferring hundreds of billions of dollars annually from military spending to green the economy and head off our real national security threat (global warming), and forcing big corporations and the wealthy to pay their fair share of taxes, we will have to pay for the GND in the same way we paid for the New Deal of the 1930s and World War II: through increased government spending and by issuing low-interest and no-interest government bonds.

With that $2 trillion available, the United States can finance our twenty-first-century New Deal and the global World War II–like campaign against climate catastrophe and runaway global warming by appropriating more money for existing programs that support the GND; creating new programs that support the GND; making loans or providing loan guarantees to taxpayers, farmers, businesses, municipalities, counties, and states to retrofit the economy for Regeneration; providing jobs for millions of unemployed/underemployed workers and youth in a nationwide Climate Conservation Corps; and paying subsidies or providing tax breaks to taxpayers, businesses, and farmers to carry out the renewable energy and regenerative food, farming, and land use practices that we need.

As Ellen Brown of the Public Banking Institute and a number of proponents of modern monetary theory (MMT) explain, the United States can easily afford to pump several trillion dollars a year into the nation's economy (similar to what Japan and China and other nations have done) in order the meet the goals of the GND and revitalize the economy—while avoiding unnecessary inflation at the same time. Brown states:

> MMT advocates say the government does not need to collect taxes before it spends. It actually creates new money in the process of spending it; and there is plenty of room in the economy for public spending before demand outstrips supply, driving up prices. . . . A network of public banks including a central bank operated as a public utility could . . . fund a U.S. Green New Deal—without raising taxes, driving up the federal debt, or inflating prices.[31]

Franklin D. Roosevelt's New Deal of 1933–41 was financed through the Reconstruction Finance Corporation (RFC), which provided funds to local and state governments and made loans to banks and businesses. The RFC got its money from the sale of bonds, but as Brown points out, "proceeds from the loans repaid the bonds, leaving the RFC with a net profit. The RFC financed roads, bridges, dams, post offices, universities, electrical power, mortgages, farms, and much more; and it funded all this while generating income for the government."[32]

Zach Carter in the *Huffington Post* summarizes the revolutionary implications of modern monetary theory:

> Modern monetary theorists believe that confusion around money has distracted economists from the real things that affect the economic health of society—natural resources, technology, available labor. Money is a tool governments use to manage these variables and solve social problems. It is not a scarce resource that governments have to track down in order to pay for projects.[33]

As Professor Stephanie Kelton, a prominent spokesperson for modern monetary theory and an economic advisor for Bernie Sanders in 2016, points out:

> A Green New Deal will actually help the economy by stimulating productivity, job growth and consumer spending, as government spending has often done. . . . The federal government can spend money on public priorities without raising revenue, and it won't wreck the nation's economy to do so. That may sound radical, but it's not. It's how the U.S. economy has been functioning for nearly half a century. That's the power of the public purse. . . . As a monopoly supplier of U.S. currency with full financial sovereignty, the federal government is not like a household or even a business. When Congress authorizes spending, it sets off a sequence of actions. Federal agencies, such as the Department of Defense or Department of Energy, enter into contracts and begin spending. As the checks go out, the government's bank— the Federal Reserve—clears the payments by crediting the seller's bank account with digital dollars. In other words, Congress can pass any budget it chooses, and our government already pays for

everything by creating new money. This is precisely how we paid for the first New Deal. The government didn't go out and collect money—by taxing and borrowing—because the economy had collapsed and no one had any money (except the oligarchs). The government hired millions of people across various New Deal programs and paid them with a massive infusion of new spending that Congress authorized in the budget. FDR didn't need to "find the money," he needed to find the votes. We can do the same for a Green New Deal.[34]

So What Do We Do Now?

It's getting late. If we're going to preserve a livable Earth, we, the global grassroots, must do more than mitigate or slightly reduce global warming. We must *reverse* global warming by eliminating greenhouse gas emissions and drawing down several hundred billion tons of excess carbon from the atmosphere. But how?

First of all, we're not going to reverse climate change by supporting moderate, middle-of-the-road candidates for political office who tell us the GND is too radical or too expensive. We're certainly not going to solve our climate emergency by opting out of politics altogether. We're not going to save ourselves by politely asking out-of-control fossil fuel and agribusiness corporations and indentured politicians to please stop destroying the planet, poisoning our food, and threatening the health and future of our children. We're going to have to organize a local-to-global grassroots rising and consciousness-raising campaign. We're going to have to create a grassroots movement capable of disrupting business as usual, carrying out a ballot box revolution, and implementing a GND, not only in North America, but across the world.

Second, we must avoid pinning our hopes for survival and climate stability on high-tech, unproven, and dangerous "solutions" such as geo-engineering (blocking sunshine, attempting to alter global weather patterns, or changing the chemical composition of the oceans), GMOs and fake meat, or carbon capture and sequestration (CCS) for coal plants. These radical engineering techniques are not only untested and unaffordable but inherently unpredictable and dangerous.[35]

Finally, we must stop naively believing that *soon* (or soon enough), several billion consumers all over the planet will spontaneously abandon their cars, air travel, air conditioning, central heating, and fossil fuel–based diets and lifestyles just in time to prevent atmospheric concentrations of greenhouse gases from moving past the tipping point of 450 ppm of CO_2 to the catastrophic point of no return. If we are going to survive, we must quickly move toward regenerative, carbon-sequestering food, farming, and land use, as well as renewable forms of energy.

Public and Private Investment

< regeneration driver four >

Ultimately, we need to transform finance and shift the flow of investment capital to perpetuate a Regenerative Economy that serves humanity and is a steward of Earth's ecosystems.

JOHN FULLERTON and HUNTER LOVINS,
"Creating a Regenerative Economy to Transform Global
Finance into a Force for Good"

Impact investments in agriculture represent a huge opportunity to implement deep changes in farming methods and adopt new ways of producing food that help mitigate climate change through lower greenhouse gases emissions, help respect the environment and biodiversity, enhance food security and climate change resilience and promote social equity.

VALORAL ADVISORS, "Impact Investing in
the Global Agriculture and Food Investment Space"

A s outlined in the previous chapter, we will need to bring about massive political change in order to fund and implement a multitrillion-dollar Green New Deal (GND). This GND will drastically increase public investment in renewable energy, regenerative land use, job creation, social services, public education, and infrastructure development, resolving our climate emergency and addressing our longstanding social, public health, and economic crises. But in addition to these huge political and public investment changes, we also need to transform private commerce and investment practices.

We need a massive shift in local, regional, and global investments away from degenerative energy, food, farming, and land use practices, as exemplified by "blue chip" investments in the stocks of the Fortune 500 corporations, to renewable and regenerative practices. Given the enormity of our world-changing endeavor, we're not going to be able to adequately fund a rapid transition to 100 percent renewable energy and regenerative practices over the next twenty-five years with just government bonds, loans, subsidies, and investments, nor with nonprofit grants, grassroots crowd-funding, and our own meager revenue or savings, no matter how hard we try. To move toward a new regenerative system will require not just government support, but massive amounts of private investment capital as well. So-called progressive foundations—for example, the Kellogg Foundation, Rockefeller Foundation, Pew Charitable Trusts, and Ford Foundation—that dole out several hundred million dollars annually in grants to nonprofit energy, food, farming, and land conservation organizations and projects, ostensibly to promote sustainability or other public interest goals, actually are donating only about 5 to 7 percent of their assets or endowments to tax-deductible organizations. The other 90-plus percent of their money, *hundreds of billions of dollars*, remains locked up in conventional stocks, bonds, and mutual funds, propping up "business as usual" for the Fortune 500, while earning, for their collaboration, 3 to 5 percent or more, on average, in interest every year.[1]

The average financial return on these foundations' destructive investments in fossil fuels and industrial food and agriculture is actually less than what they could earn, on average, by investing in local organic and regenerative farms and small businesses—enterprises that generate far more jobs and economic benefits for everyday citizens, instead of simply fattening the coffers of the wealthy. As Michael Shuman puts it in his book *Local Dollars, Local Sense*, conventional investors are "overinvesting in Wall Street and underinvesting in Main Street."[2] The supposedly progressive, environmentally concerned fund managers for these foundations, along with pension fund managers and bankers, justify keeping most of their assets in conventional stocks, bonds, and mutual funds as "prudent and natural" (i.e., everybody does it) in order to guarantee their fund's solvency or, in the case of foundations, their ability to continue to give out donations on a long-term basis. This degenerative business as usual must change if we're going to move forward.

Yes, public interest organizations, such as those that I help manage, need to keep on raising money through foundation grants and donations from our members, as well as revenue generated by our farms or businesses. But we're also going to have to tap into the massive capital resources of the progressive- and even not-so-progressive-minded investors who control the bulk of the United States' $30 trillion in liquid assets, but who also increasingly share our desire to save the planet from catastrophe.[3] To divest from Degeneration and reinvest in Regeneration, we're going to have to educate not only consumers and policy makers but investors as well, pointing out and demonstrating in financial and business terms that our organic and regenerative pilot projects, current best practices, and plans for increasing market demand and scaled-up production are not only necessary but investment-worthy. We're going to have to show "impact investors," as many now call themselves, that strategic regenerative enterprises, infrastructure, operating capital, and management can advance our common regenerative goals while at the same time giving them a fair return on their investment, either slightly below or roughly comparable to what they are earning now.

Regeneration as Entrepreneurship

At the present time, trillions of dollars are invested in the industrial, so-called "conventional" global food system, a system with annual sales to consumers of $7.55 trillion.[4] This system is artificially propped up by hundreds of billions of dollars in annual government subsidies, sustaining energy-intensive, chemical-intensive, GMO-driven agriculture and food processing. The massive external costs and damages of the conventional food system—that is, degeneration of the environment, public health, and climate stability—are hidden from citizens, taxpayers, and investors alike (in other words, those who will eventually pay the bill for this collateral damage), with "true costs" never included in the bottom line. If true cost accounting were required, massive amounts of capital would move out of industrial food, farming, and land use into more productive sectors such as organics, holistic grazing, agroforestry, and agroecology. Unfortunately, very little money, relatively speaking, is currently invested in the organic, agroecological, and regenerative sector. According to one recent study, there is already approximately $321 billion invested in regenerative, "soil wealth" enterprises and projects—but that amounts to just a little over 1 percent of current global liquid assets.[5]

Meanwhile, investors, pension managers, mutual funds, sovereign funds, and even so-called green or socially responsible funds that have begun to shun fossil fuel investments continue to pour money into climate-degrading, chemical-intensive food, factory farm, and commodity corporations, GMOs, patented seeds, and most recently "Big Data." In the agribusiness market "Big Data" includes high-tech tools of soil, climate, weather, crop, markets, and chemical inputs analysis, utilizing computers, satellite data, drone photography, sophisticated farm machinery and other techniques to supposedly increase yields and profits on modern farms.

How do those of us who are not fundamentally business-minded nor trained in soliciting investments become more adept and sophisticated in obtaining funding to scale up our regenerative economy? How do we prove to investors that there is a strategic, indeed existential imperative to consider investments that not only are socially responsible and good for the environment but also build the essential new regenerative economy?

Growing up in a decidedly working-class family, spending summers on my grandparents' small family farm, and then waging campaigns for most of my adult life against corporations like Monsanto and other "bad guys in suits," I must admit that it has been a stretch for me to finally accept the fact that businesspeople and investors will have to become important drivers of Regeneration. I've always felt somewhat uncomfortable around the East and West Coast MBA types who run the foundations that give out money, if you're lucky, to public interest activist groups like the Organic Consumers Association. I feel out of place in corporate offices and settings. I feel strange in a suit and tie. I feel more comfortable in the Minnesota North Woods, or at our high desert farm school in Mexico, than I do in the city. Our Minnesota office and agroecology farm is located in a town called Finland, population three hundred. Six miles away, I live with my wife and my son in a cabin, yurt, and straw-bale office in the middle of the woods; for our first ten years here, we had no running water. The closest village to our organic farm and conference center in Mexico is called Membrillo, and it's too small to even be on the map.

But the time has come for old hippie back-to-the-landers like myself, and for everyone else who cares about the climate and the survival of civilization, to not only step up our campaigning but become more entrepreneurial in our thinking and activism as well. After spending decades struggling to make ends meet while working in, and later directing, financially strapped not-for-profit organizations, cooperatives, activist campaigns, and a number of organic small farms and businesses, until recently I'd never sat down with a real investor, as opposed to a philanthropist (someone from whom I was asking for a tax-deductible donation), and talked about why they should invest money in a regenerative enterprise.

Despite writing hundreds of grant applications and raising millions of dollars every year—usually the hard way, via $25 to $50 donations from members and supporters—until recently it never occurred to me to solicit investors for a business venture separate from the nonprofit sector that could advance the regenerative economy. It never occurred to me to approach the handful of large donors who have always supported organics to become investors in a common endeavor that could advance our common concerns and still generate a fair return to them as investors. But recently I've started moving in this more entrepreneurial direction, as have a growing number of others in our global Regeneration movement.

If we want to change the world and reverse climate change, many of us have no choice but to expand our activist horizons and take on the role of entrepreneurs, food and land use systems designers, business planners, project incubators, marketers, and fundraisers. If we don't presently have these skills—and of course most of us don't—we need to reach out to friends and supporters who do have these business and financial management skills and get them to help us. Without millions, indeed billions, of dollars invested in scaling up regenerative food and farming enterprises, production and marketing co-ops, and businesses, we will fail. We frankly don't have time to convert the world to regenerative thinking and practices over the next hundred years, one consumer, one farmer, one retailer at a time. We have to speed up the process, as our Regeneration International banner at the UN's global climate summit in Germany in 2017 proclaimed: "Speed Up the Cooldown!"

Combating Degeneration: A Global Industrial Agriculture Divestment Campaign

By now you've no doubt heard about the ongoing and highly successful global campaign waged by climate change activists to get institutional investors and fund managers, including churches, universities, state pension funds, cities, and even the massive Norwegian sovereign wealth fund (with $900 billion in assets) to divest from fossil fuel corporations.[6] This modern grassroots-powered divestment movement, based in part upon the successful anti-apartheid South Africa divestment movement of the 1980s, as well as the more recent anti–Big Bank "Move Your Money" campaign in the United States, has educated a wide section of the body politic (including a growing number of socially responsible investors) to understand that not only is fossil-fuel-driven climate change a serious threat to our common survival, and therefore morally indefensible, but it is actually a bad investment in financial terms, given that trillions of dollars of fossil fuel reserves in the ground (coal, oil, natural gas, and uranium), once the climate crisis gets worse, will become "stranded assets," never to be excavated or burned.

So far, in just a few years, the fossil fuel divestment movement has managed to move over $6 trillion in assets out of the coal, petroleum, and nuclear sector, with a significant amount ($300 billion annually) being reinvested into solar, wind, and energy conservation efforts. As this movement continues to gather momentum (and as renewable energy continues to become cheaper and more profitable than dirty coal, oil, and fracking), more and more institutional investors will undoubtedly continue to move their money, eventually forcing even the fossil fuel multinationals themselves to remove stranded assets from their balance sheets and move to renewable forms of energy, or else go bankrupt.

What organizations like 350.org and student, church, and other grassroots climate activists in the divestment movement have done in the energy sector, discrediting degenerative fossil fuel energy corporations and practices and calling for financial divestments, we now need to do in the food, farming, and land use sector. Our strongest arguments are that global industrial food, farming, and land use practices (chemical- and energy-intensive farm inputs and production, processing, packaging, refrigeration, transportation, deforestation, and waste) are generating a full 43 to 57 percent of all current greenhouse gas (GHG) emissions—an amount roughly equal to all the emissions in the nonfood transportation, utilities, building, and manufacturing sectors. These GHG emissions are fueling global warming and ever more severe climate change. In addition, conventional degenerative food, farming, and land use, exemplified by the practices of the Fortune 500 corporations (Bayer/Monsanto, Walmart, McDonald's, Cargill) that dominate or finance this sector, are destroying our environment, our health, biodiversity, and the livelihoods of the world's three billion farmers, herders, forest dwellers, fishing communities, and rural villagers.

As our counterparts in the fossil fuel divestment movement have pointed out, not only are degenerative food, farming, and land use investments immoral, in the sense of threatening our very survival, but they are actually bad investments, once you tally up the $8.5 trillion annual "true costs" and collateral damage to our health, environment, and social fabric.[7] The thousands of multimillion-dollar lawsuits now being filed against Monsanto (for its cancer-causing pesticides) and DuPont (for its Teflon and PFOA/PFOS chemicals) for knowingly and deliberately poisoning millions of people as well as our water and environment (like the anti-tobacco lawsuits of the 1990s), are just a preview of the "judgment day" coming for transnational food, farming, timber, and extraction corporations unless they admit the deadly error of their ways and turn toward regenerative practices. And, of course, as the climate change divestment movement expands its focus from fossil fuels into the food and farming sector, it will become easier to make the case for reinvestment in the $100 billion to $500 billion organic, agroecological, and regenerative food, farming, and land use sectors.

Since the 1960s and '70s, reacting to mounting grassroots pressure and damage to their companies' and brands' credibility, reputations, and bottom lines, large corporations have adopted, with varying degrees of sincerity and success, the concepts of corporate social responsibility (CSR) and socially responsible investment (SRI). Basically, proponents of CSR and SRI claim that it is possible to generate significant profits (both short-term and long-term) for their executives and shareholders while at the same time "doing good things" for the environment, their workers, and the general public. Sometimes referred to as the "triple bottom line" (that is, profits, people, and the environment), CSR has now been embraced, at least rhetorically, if not substantively, by nearly every major corporation in the world, as well as many small and medium-sized enterprises. In terms of investment, SRI fund managers in the United States now claim that over $2 trillion of the nation's $30 trillion in liquid assets are invested in a socially responsible manner, earning an adequate return but screening out unethical corporations.[8]

Although no one would argue that CSR and SRI concepts are a bad idea, the question is whether these idealistic concepts are actually being put into practice in a manner that really makes a difference, or whether they are being used more as a public relations scheme, a type of greenwashing to cover up and perpetuate corporate greed, misdeeds, and degenerative practices. Unfortunately, as critics such as Michael Shuman and Paul Hawken have pointed out, most SRI investments are no better than non-SRI investments. SRI fund managers may indeed, at your request, screen tobacco companies or armaments manufacturers from your personal portfolio, but the multicorporate, multinational, blue chip mutual funds they put your money in are still the same Fortune 500 fossil fuel companies, corporate agribusiness giants, and Big Pharma companies wreaking havoc on public health, the environment, and climate. In other words, SRI investment schemes are just another type of greenwashing, providing the illusion of change while actually propping up the status quo. Genuine SRI investments, on the other hand, would focus on solving the most serious problem we face—global warming—by divesting from industrial agriculture and fossil fuels, while reinvesting in renewable and regenerative enterprises and practices.

Today's large corporations typically have a CSR department and staff, and usually an outside public relations firm, or several, to handle press relations, to maintain their image, and to work in coordination with other corporate departments, including human (employee) relations, advertising, social media, marketing, and customer relations. Glossy CSR reports are printed up every year highlighting the corporation's good deeds and contributions to society and the environment. Unfortunately, most CSR, like SRI, is propaganda. If real CSR was a reality for the large corporations that dominate the global economy, it's unlikely that greenhouse gas emissions would still be rising, that almost a billion people would still be going hungry, that chronic diseases (including food- and environmentally caused cancers, heart disease, obesity, birth defects, and behavioral disorders) would be reaching epidemic proportions, or that air, water, toxins, and plastic pollution would be getting worse, not better.

The fundamental problem is that there are no strict definitions, standards, or independent third-party verification of CSR or SRI claims, in contrast to more reliable standards, such as certified organic, grass-fed, or Fair Trade standards in the food or clothing sector, or Forest Stewardship Council standards in the timber and wood products sector. When corporations operate under vague, voluntary CSR standards and fail to use true cost accounting, profits inevitably trump social responsibility. Publicly traded corporations (as opposed to family-owned, privately held, or not-for-profit enterprises) have a fiduciary duty and legal responsibility to maximize profits to their stockholders, not just in the long run, but especially, in the United States, in the short run as well. If they don't maximize profits, they can be sued by their shareholders. (And by "short run," we're talking about sales and profits for the last quarter, or three-month period, with no thought for the long-term impacts of company policies and practices on the climate, the environment, and future generations.) Any CEO or board of directors from a publicly traded corporation that dared to put environmental health, climate health, or public health above short-term profits would not last very long.

Today's corporate balance sheets do not typically include the externalities or collateral damage of corporate business as usual on human health, the environment, and the climate. Consequently, Bayer/Monsanto, Dow, DuPont, or Syngenta/ChemChina may indeed have installed solar panels on their corporate offices last year, or they may have raised the salaries of their workers a little bit, but the bottom line is that they sell GMOs and toxic chemicals that damage soil, human health, and the climate as their primary money makers. Their products cause billions of dollars of damage to public health, soil health, the environment, and climate stability every year. These multibillion-dollar externalities do not show up in the corporations' balance sheets (unless they happen to get sued and are forced to pay out damages), nor in their annual CSR reports. Similarly, Boeing or Lockheed Martin may have turned a tidy profit this quarter and may have an excellent recycling program, but their profits basically depend on building fossil-fuel-guzzling commercial airplanes that are destroying the climate, along with warplanes, missiles, and bombs that are killing people and obliterating cities and towns. Similar constraints shackle pension fund managers, who, under law and custom in the United States and other nations, must strive to maximize returns on investments, with no consideration of collateral damage.

Even the term "regeneration" is now being bandied about by corporations to tout their CSR practices. If we want to preserve the integrity of our Regeneration movement, we need a clear and transparent definition of what regenerative and transitioning-to-regenerative food, farming, and land use actually is. (In my view, it means no toxic synthetic chemicals, GMOs, fertilizers, or animal drugs allowed, with annual and continuous increases in sequestered soil carbon and biodiversity.)[9] As the climate crisis intensifies and the demand for regenerative practices and products increases, we can expect more and more large (and mainly conventional) corporations such as General Mills, Unilever, Danone, and others to claim that they are embarking on the road to Regeneration, meanwhile continuing to allow the use of chemical fertilizers, pesticides, new-generation GMOs (with gene-edited RNA), and proprietary hybrid seeds to produce the bulk of their products, many of which are nothing more than junk foods.

Major corporations, international funding organizations, and billionaire investors like Bill Gates and Blackrock are now stepping forward with new, supposedly climate-friendly food and farming financial schemes and technology packages such as "climate-smart agriculture" and "zero budget natural farming."[10] Although investment capital for genuinely regenerative food, farming, and ecosystem restoration projects is scarce right now, we should anticipate that big corporations are soon planning to jump on the bandwagon, with their own versions of climate-smart or natural farming, just as soon as regenerative thinking and practices start to achieve critical mass in the marketplace.

Fair-return capital investments with no strings attached will always be welcome in our new regenerative economy. But we must beware of big corporations and big investors claiming that we can move to regenerative practices by utilizing billions of dollars in high-interest loans to cash-strapped farmers, supplying them in turn with GMOs, proprietary seed, and expensive Big Data schemes, while sidestepping traditional certified organic and agroecological practices for local and regional markets.[11] Globally there are approximately 570 million farms, of which 2.7 million are certified organic, while a much greater number, approximately 25 to 50 million farms, are using agroecological (organic but not certified organic) practices. These are the types of farmers and farming practices that need to be supported—not those that need the support of Big Chem.

We've gone from approximately $1 billion in certified organic sales in 1990 to a current global market share of $100 billion, with an army of small agroecological farmers likely producing probably ten times as much for local consumption and local markets. The fundamental challenge then becomes, how do we scale up these best practices in organics and agroecology over the next few decades? How do we move today's organic and agroecological farmers, along with all the hundreds of millions of other farmers, some of whom are struggling just to survive, to the next stage—that is, to regenerative food, farming, and land use? Although we don't have the space here to go into great detail on the investments we will need, the following are several strategic areas where we require immediate action.

Examples of Priority Areas for Regenerative Investments

As we've emphasized in earlier chapters, it is the multitrillion-dollar global factory-farm system, interlocked with the network of chemical-intensive monoculture grain production, that is a primary driver of the present degeneration. If we can replace this factory-farm system with one based upon holistic, pasture-based, organic animal raising and multispecies, perennialized, no-till, cover-cropped, agroforestry grain production, we will be well on our way to solving our most serious crisis. How do we do this? Specifically, what kind of investments in terms of land, land restoration, livestock, infrastructure, and marketing do we need to speed up this process, propelling regenerative farming and ranching into a trillion-dollar powerhouse over the next twenty-five years, whereby regenerative practices become the norm, rather than just the alternative to factory farming and industrial (chemical- and GMO-intensive) grain production?

First, if we want healthier soil, more carbon sequestration, biodiversity, nutrient-dense food, healthier animals, rural prosperity, and a livable environment, we have to move fifty billion farm animals out of intensive-confinement factory farms and feedlots and put them back out on the land, grazing and foraging in biodiverse pastures (with multispecies plants, grasses, and trees) in a manner that increases soil fertility, water retention, and aboveground crop and biomass growth. We have to reduce the massive acreage now devoted to monoculture GMO corn and soybeans and restore these fields to what they once were: biodiverse, multispecies grasslands, preferably silvopasture or agroforesty grasslands with massive carbon sequestration potential. Part of how we can do this, as indicated earlier, is to educate consumers to boycott factory-farmed foods and pay a fair price to farmers who are "doing the right thing," utilizing organic and regenerative practices. We also need to get farmers, ranchers, agriculture schools, and extension services to share and promote best practices for grazing, animal husbandry, and organic grain growing, as well as lobby policy makers to support regenerative farming. In addition, we need significant new investment in land acquisition, livestock for holistic grazing, farmland and pasture improvement, processing facilities, transportation and storage infrastructure, and marketing. One major step toward disrupting the degenerative factory-farm paradigm is to move as quickly as possible to grass-fed beef, lamb, goat, and buffalo.

The US beef market is huge, at $80 billion a year, with over ninety million head of cattle, one-third of which are slaughtered every year.[12] The good news is that there is already an enormous consumer demand for grass-fed/grass-finished beef. In the United States alone, 100 percent grass-fed beef, with little or no government support, currently generates $5 billion a year in sales, with $1 billion produced domestically by more than six thousand ranchers, and the rest imported. Millions of consumers, restaurant chefs, and food retailers already understand that grass-fed beef is more nutritious and healthy; better for the environment, carbon sequestering, and climate; free of pesticides, artificial hormones, and animal drug contaminants; better-tasting and flavorful; and more humane for the animals. This is why millions of consumers are willing to pay a premium price for 100 percent grass-fed.

The bad news is that US ranchers' profit margin for grass-fed beef is very small, causing grass-fed products to be priced significantly higher than US feedlot beef and imported grass-fed products. A major reason for this price differential is that small grass-fed ranchers are charged much more for processing their cattle than the large feedlot aggregators (like JBS, Tyson, Cargill, and National Beef). In addition, small producers incur additional costs because they often have to transport their animals over long distances for processing at USDA-inspected plants. As a result, most (75 to 90 percent) of the grass-fed, grass-finished beef being consumed in the United States is imported from overseas, from Australia, New Zealand, Uruguay, and Argentina. Meanwhile the overwhelming majority (95 percent) of US cattle (after grazing on grass for twelve to eighteen months) are sent to hellish factory-farm feedlots. There they are fattened up on (nonorganic) commodity grain, growth hormones, and antibiotics before being sent to massive slaughterhouses, where filth and disease are rampant and labor exploitation is the norm. The low-grade, often contaminated cheap meat from these factory-farm feedlots is sold at a cheaper price than grass-fed beef, but factory-farmed, grain-finished beef has enormous hidden costs in terms of damage to human health, the environment, animal welfare, and the climate.

Several types of investments are needed in order expand the market and profitability of grass-fed beef (as well as grass-fed dairy products, grass-fed cattle by-products, and organic pastured poultry and pork), not just in the United States, but in most nations:

Infrastructure. Significant investments are needed to build or rebuild local and regional infrastructure for small regenerative ranchers and poultry farmers, including local "custom" processing plants (usually state-inspected, rather than USDA-inspected) for meat production in every region. To be profitable, small meat and dairy farmers also need to gain income from all parts of the cow or animal, not just the meat, including bones and hides. This will require investment in regional processing and dehydration plants to produce high-value beef and dairy by-products such as organic/grass-fed powdered bone broth, whey protein powder, and ingredients for nutritional supplements. In addition, the grass-fed market needs new ecologically friendly tanning and production facilities to manufacture grass-fed leather products, which can successfully compete in and disrupt the $90 billion global leather market.

Land acquisition. Investments are needed for regenerative ranchers or would-be ranchers (and organic grain producers) to buy or rent pasture- and cropland. One example of a successful financial enterprise facilitating land acquisition for organic and transitioning-to-organic farmers is Iroquois Valley Farmland, which purchases farmland for new or experienced farmers who want to farm organically and rents this land out to farmers for seven to nine years, after which they have an option to buy the land. Another successful example of regenerative investments is Farmland LP, which has shown the profitability of buying out conventional chemical farms and converting them to organic.[13]

Land and pasture improvement. In order to maximize soil and climate benefits, a sufficient number of cattle are required in order to avoid undergrazing as well as to achieve a scale of production that is profitable. Finishing ranches must be developed where cattle can be aggregated together in larger herds ("mob grazing") and grazed holistically to fatten them up before slaughter. These finishing ranches need to have high-quality pasture grass and tree cover, as do all grazing operations, including those with sheep, goats, and other herbivores. Investment funds to purchase land or additional livestock and to make pasture improvements are all-important in this regard. In all of these areas, regenerative producers and advocates need to connect with impact investors to identify investment opportunities and make deals.

Marketing co-ops (advertising and wholesale) **and marketing infrastructure** (transportation and storage) are also necessary for the growth of the grass-fed and regenerative sector. Once a critical mass of consumers become fully informed about the horrors of factory farms and have access to nutritionally superior/climate-friendly regenerative meat, dairy, and poultry (priced fairly but not exorbitantly), the factory-farm system (and the monoculture grain system that props it up) will collapse. As the market demand for cheap meat, dairy, and poultry shrinks, we will see major synergistic growth in organic and regenerative grain production, which is necessary in order to provide high-quality animal feed for regenerative pork and poultry operations. (Pigs and poultry, in contrast to herbivore cattle, sheep, buffalo, and goats, are naturally omnivores and typically need grain in their diets in order to be healthy—and profitable for the farmer.)

Other Priority Sectors with Major Investment Potential

The organic and regenerative food and products sector, including clothing, body care, and medicinals, has grown fiftyfold over the past three decades and will likely continue to grow rapidly over the next few decades as health, environmental, and climate concerns become ever more important. A number of new food, fiber, or beverage sectors with tremendous (multibillion-dollar) potential include organic craft or artisan beer, organic cannabis and industrial hemp, and urban organic agriculture. The multibillion-dollar global ecotourism sector is another area where organic and regenerative foods and products can become the norm.

The retail dollar sales of craft beer in the United States in 2017 increased 8 percent from the previous year, up to $26 billion, amounting to more than 23 percent of the $111.4 billion US beer market.[14] Globally, craft beer sales are projected to reach $100 billion by 2025. Unfortunately, very little craft beer (which typically utilizes malted grains such as barley and wheat as well as hops) is currently organic, much less regenerative, while only a small percentage of craft brewers have branched out into brewing with nongluten (and at times perennial) grains such as millets, sorghum, buckwheat, quinoa, and amaranth.[15] The growth of organic, nonbarley, and nonwheat beers is likely to change, however, as consumers learn about the widespread contamination of beer (including craft beer) with pesticide residues and the impact of nonorganic grain growing on the environment and climate, and as gluten intolerance becomes even more of a health issue. Anheuser-Busch has already come out with a low-alcohol organic beer as well as a multimillion-dollar program to help barley farmers make the transition to organic. And, of course, the growth and increasing biodiversity in the organic grain and beverage sector will be beneficial to the growth, availability, and cost of animal feed in the organic and regenerative poultry and pork sectors as well.

The growing market for legal cannabis, CBD, and industrial hemp is another agricultural sector with tremendous potential for organic and regenerative production. By 2025, potential sales of legal cannabis are projected to reach $25 billion in the United States and hundreds of billions of dollars around the world.[16] Many small organic farmers are now seeing that even small plots of high-grade organic cannabis can serve as a lucrative cash crop to enable them to grow lower-profit organic produce, fruits, and grains and raise livestock on a biodiverse farm while still turning a profit.

Mapping, Network Building, and Regional Investment Planning

Regeneration activists are taking a number of movement-building and enterprise-building actions across the world to prepare the ground for attracting not only grassroots and philanthropic donations, but investment capital as well. One important first step is *mapping*—that is, identifying and highlighting best organic and regenerative practices and practitioners in each area, including organic and agroecological farmers, ranchers, land managers, retail outlets, restaurants, farmers markets, community-supported agriculture (CSA) networks, food hubs, foodsheds (regional food production and distribution systems), and consumer/community education programs (gardening, cooking, nutrition, school education, et cetera). Utilizing preexisting databases and directories, followed up by direct contact (e-mail, telephone, site visits), it is possible to identify the best organic and regenerative producers, retailers, processors, educators, public policies, policy makers, educators, and public interest organizations in each area. Newsletters, listservs, social media, conference calls, and local/regional gatherings are facilitated through this mapping. Basically the goal of Regeneration networks is to figure out how to replicate and scale up the best practices in each region.

A second step, now under way in many areas, is to follow up with *network building*—that is, contacting regenerative producers, processors, retailers, and businesses and bringing them together into a market network, coalition, co-op, trade group, or other market-oriented enterprise, where producers can share best practices, identify major obstacles to growth, and develop plans for generating greater market demand, achieving economies of scale, reforming public policy, and attracting additional capital for necessary infrastructure expansion (processing, storage, transportation, distribution). On a parallel track, again following up on mapping, regional regenerators can build consumer and public education Regeneration networks to increase market demand, crowd-fund organic and regenerative enterprises, and lobby for policy change and needed public/private financing, such as supporting the GND.

A third step, as outlined earlier, is to *build coalitions* with like-minded public interest groups such as climate change groups, student groups, environmental organizations, progressive political groups, church organizations, peace groups, natural health advocates, progressive and green businesses, and others—again, to lobby for policy change and to connect the dots between climate change and other hot-button issues such as renewable energy and green jobs.

Finally, as networks and coalitions coalesce, they become ready for a fourth step, which calls for local to regional long-range planning and strategizing on how to strengthen and scale up all four major drivers of Regeneration: consumer demand, farmer innovation, public policy change, and capital investments. Once these components are in place, it should be possible to catalyze even larger regional and even national Regeneration networks and to begin to focus on international networking as well.

Investment Structures and Strategies

Attracting a critical mass of investment capital into the regenerative sector requires at least three components: a group of investors who have a pool of money and are willing to invest; a financial intermediary corporation (fund manager) that can receive capital investments and make loans and provide supervision for loans to investment-worthy enterprises or co-ops; and a network of well-organized regenerative enterprises or businesses. It is obvious that there are a growing number of impact investors out there who are willing to invest in viable businesses that are regenerative and climate-friendly, given the increasingly serious nature of our environmental, public health, and climate crises. These impact investors simply need to be approached in the right way by the right people. The problem is that these investors typically don't want to directly oversee or manage a loan to a particular farm (or set of farms), production facility, or processing facility themselves. They need an intermediary or fund manager with a proven track record who can administer the loans and guarantee a fair return on their investment. In turn, these intermediaries or fund managers need enterprises (farms, co-ops, production facilities, new product brands) that can demonstrate their ability to succeed, with proof of market demand (hopefully including purchase orders or provisional contracts), proof of experience and capabilities, proof of concept (pointing to preexisting, successful models), a well-thought-out multiyear business plan, and, if possible, some sort of collateral or cosigner for loans. In this regard, it is important to come up with business plans and investment opportunities that, rather than benefiting just one farm, food business, or enterprise, will strengthen and expand the market for a whole system or network of producers, processors, and sellers.

One example of a successful financial intermediary corporation that is helping to expand the organic and regenerative sector is Iroquois Valley Farmland, mentioned earlier, a company based in the Midwest that solicits investments from impact investors and then buys farmland for certified organic or transitioning-to-organic farmers. Farmers lease the purchased farmland for a period of seven to nine years (this rent pays a return to the original investors), with an option to buy the land at the end of the lease period after they have successfully established an organic farm and business (and meanwhile have improved the land). With a proven track record, after seven to nine years, these farmers can then get a regular mortgage on the land from a bank or credit union, freeing up the original capital investment by Iroquois Valley to reinvest in more organic farmland. Should the farm fail during the initial lease period, Iroquois Valley would still have the land.

Another example from the Midwest is Shared Capital Cooperative, a financial intermediary and fund manager that makes loans to producer or marketing co-ops.

International Investments in Regeneration

As emphasized throughout this book, the Regeneration revolution that we need is global in scope. Without regenerating billions of acres of cropland, pastureland, forest, and even urban landscapes across the world, especially in the tropical and semitropical areas of the developing world, where soils are most degraded and where soils and forests have the greatest capacity to relatively quickly sequester billions of tons of excess carbon from the atmosphere, we will not be able to mitigate, much less reverse, global warming. No matter what we do in terms of regenerating our food, farming, and land use practices in North America and Europe, it will not be enough to forestall runaway global warming and climate catastrophe without fundamental change in the Global South as well.

We need major regenerative grants, loans, and investments in Asia, Africa, and Latin America (the Global South), as well as in the Global North. Our staff and affiliates from Regeneration International have heard from organic and agroecological farmers and livestock managers from all over the world that international funds (like those of the World Bank, UN, FAO, USAID, Global Environment Framework, Green Climate Fund, and affiliated national government funds) are very difficult to access. These funds, whether in the form of subsidies, grants, or technical assistance, are routinely hijacked by larger, well-connected farmers or siphoned off by international bureaucrats and corrupt national and local governments, rather than getting into the hands of small farmers and rural communities, especially those engaged in organic and agroecological practices. The model that needs to be implemented instead is one whereby best practices and practitioners are identified, and then their efforts are replicated and scaled up with adequate funding and technical and marketing assistance.

One prime sector ripe for investing is that of the holistic management hubs practicing regenerative grazing in the United States, Latin America, Zimbabwe, Zambia, and a dozen other nations. These centers can provide positive training and examples for hundreds and even thousands of ranchers to demonstrate how grass-fed beef, grass-fed dairy, and holistic livestock management can be not only profitable and good for the environment and wildlife but also important drivers of carbon sequestration in soils, including several billion acres of degraded pasturelands across the world. The same goes for diversified organic vegetable and grain production and farming practices that include agroforestry and perennial crops.

Farmers, ranchers, herders, forest dwellers, and urban agriculturists in every region of the world have already developed organic and agroecological practices (whether certified or not) that are regenerating soils and land-scapes. The challenge is to get the funds, trainers, and technical assistance to scale up these best practices, and to do so as quickly as possible.

Maintaining business as usual in terms of investment or foreign aid and development is a recipe for disaster, given the seriousness of our crisis. We need a new wave of regenerative investment to complement public funds, market demand, farmer innovation, and progressive policy change, and we need it now.

7

The Global Road
to Regeneration

*Just transitioning 10–20% of agricultural production to best
practice regenerative systems will sequester enough CO_2 to reverse
climate change and restore the global climate. Regenerative Agriculture can change agriculture from being a major contributor to
climate change to becoming a major solution.*

ANDRÉ LEU, "Reversing Climate Change
with Regenerative Agriculture"

In previous chapters we've outlined the disastrous and frightening
manifestations of, as well as the positive and hopeful solutions to, our
global crisis. In 2018, the world's top climate scientists announced a
new timeline—in fact, an ultimatum, based upon the latest observations
and data—on how quickly we have to drastically reduce net greenhouse
gas emissions (at least 45 percent) over the next decade, from 2020 to
2030, to keep the planet from rising above the danger zone of a 1.5°C
(2.7°F) temperature rise.[1] Unfortunately, on a hot day in 2019, we already
reached a maximum 1.2°C (2.16°F) rise in average global temperatures.

To move fast enough to meet our deadline for survival, we must make
the transition to a non-fossil-fuel, renewable-energy-based global economy
as soon as possible, but also, at the same time, we must qualitatively increase
global photosynthesis and carbon sequestration in soils, forests, and plants
so as to achieve not only net zero emissions by 2030, but significant *negative* net emissions by 2050. This dual strategy will enable us not only to
avert catastrophe but to begin to *reverse* global warming and begin the
long-term process of restabilizing the climate. This Regeneration and green
energy revolution will bring CO_2 levels in the atmosphere back down from
a dangerous peak of 450 ppm or more (which, if we don't change course,

could come as soon as 2040) to approximately 280 to 300 ppm, similar to the "safe" level of atmospheric greenhouse gases that existed for hundreds of thousands of years prior to the advent of the industrial revolution in 1750.

The bad news is that after decades of unheeded warnings and interminable delays in shifting course, we can no longer save ourselves from plunging over the climate cliff by just moving to 100 percent renewable energy. Even if, miraculously, we are able to retrofit our entire global energy, transportation, utilities, manufacturing, and housing infrastructure in the few decades we have left before we pass the point of no return, to safeguard a livable planet we must fundamentally change our agriculture, consumer, and land use practices as well. These agronomic and marketplace transformations, in turn, will only be possible if they are supported and financed as part of a system-wide political transformation, or what is now being called in the United States a Green New Deal (GND). But a paradigm shift on the scale of a GND, in political power, public policy, commerce, investment, and international relations, will only be possible if we can raise awareness on a mass scale and build powerful grassroots networks and coalitions, in every local area, region, and nation.

Accelerating Drivers

To avert disaster and restabilize the global climate, all four drivers of Regeneration will have to be operating at maximum power and synergy over the next three decades and beyond, especially in the wealthy and geopolitically influential areas of North America and Europe, but then gathering momentum in all nations, regions, and local areas. These four drivers, as discussed earlier, include:

1. Mass-based grassroots awareness, generating pressure for fundamental political and policy change, as well as a massive marketplace demand and economic stimulus for green and regenerative products.
2. Scaling up farmer, rancher, and land-management regenerative best practices worldwide so that a critical mass of Earth's twenty-two billion acres of croplands, rangelands, wetlands, forests, and marine ecosystems, as well as several billion acres of degraded wastelands, can sequester 10 to 20 Gt of carbon per year (73.4 Gt of CO_2e).

3. Mass movement and coalition building, transforming political power and public policy, along the lines of a global GND, to make the full and just transition to a green economy of 100 percent renewable energy and regenerative food, farming, and land use.

4. Transforming destructive public and private investment and business practices in order to move trillions of dollars away from fossil fuels and destructive food, agriculture, and land use practices and into green and regenerative investments and enterprises.

Of these four drivers, mass-based grassroots awareness and gaining political power are particularly key, given our limited time frame for action. Without supercharged grassroots awareness and mass mobilization against the threat of runaway global warming, we will fail. Without a US and global mobilization on the scale of the campaign against fascism in World War II, we will not be able to muster the economic and institutional resources we need to win the war against global warming and societal meltdown. And, of course, mass consumer awareness and demand complementing a US, EU, and global GND will help us achieve the changes we desperately need in commerce and the marketplace. To successfully carry out this global campaign, we will need to rapidly scale up farmer innovation and investment in the regenerative sector. But this scaling-up/drawdown process will only move forward quickly enough if we can gain political power and marshal the trillions of dollars in public and private funds we will need over the next decade and beyond to implement a transformative GND.

A Global Strategy

To reverse global warming, China, India, Russia, Japan, North America, Europe, Brazil, Indonesia, and all the industrialized and developing nations will need to accelerate their current momentum toward zero fossil fuel emissions by 2050, as pledged at the Paris climate summit in 2015. But now, in line with the warnings of global climate scientists, we have no choice but to move faster than the thirty-five-year time frame outlined in Paris and to spread the renewable energy and agriculture revolution to all corners of the globe, ASAP. The life-or-death challenge moving forward is how can we most rapidly drive the Regeneration and GND movements in the United States, Europe, and the rest of the Global North, while providing hundreds of billions of dollars in support for the developing nations of Asia, Africa, and Latin America to help our southern neighbors scale up renewable energy and regenerative food, farming, and land use in the Global South exponentially. This North-to-South aid is especially strategic in the billions of acres in the world's tropical and subtropical areas, where plant and forest photosynthesis can be ramped up most quickly, and where degraded soils in forests and grasslands can be regenerated so as to hold and store massive amounts of carbon.

Mass Consciousness Raising: Global Solidarity

The road to Regeneration is an international road, spanning every continent, country, and region, every soil type, climate, and ecosystem. Since the end of World War II and the founding of the United Nations—the last time the world cooperated on a global scale for a common cause—countries and politicians have unfortunately once again turned away from solidarity and cooperation to nationalism, hypercapitalist competition, and militarism, just as they did between World War I and World War II. Unscrupulous politicians, dictators, and corporations understand, as they have for centuries, that they can manipulate people's fears and ignorance and their cultural, ethnic, and socio-economic differences to gain and hold power and maintain their class privileges and obscene profits. But now, once again, as in the Nazi era, the world faces a common threat and a common enemy. It does no good to simply sit back and blame China, India, the United States, and Russia (the world's largest greenhouse gas polluters) for their emissions, or to focus exclusively on countries like Brazil and Indonesia for their rampant deforestation. In order to survive, we must stop denigrating and demonizing one another, put aside our differences, and start conversations and negotiations on how we can build a global green economy together, starting with renewable energy, organic food and farming, and regenerative best practices that are already being carried out in every nation of the world. Of course, wherever dictatorial regimes that deny the dangers of climate change have seized power, as in Brazil, we must support grassroots struggles to get rid of them.

The most powerful synergy on Earth, in terms of global consciousness raising and action, is to spread the positive message that we, the global grassroots, can reverse global warming and revitalize society. Utilizing the same renewable energy and regenerative practices that we need to reverse environmental destruction and create climate stability, we can restore our health (both mental and physical), improve our economic situation, and improve the quality of our daily lives. Once we start to achieve critical mass, this transcendental hope and awareness will stimulate not just one but all four of the Regeneration drivers we need to move forward. But in order to move from Degeneration to Regeneration, we must get the local, national, and global body politic to understand and truly empathize with the fact that we are all in this together—all 195 nations and the more than seven billion diverse people who live and work in our million or more local communities.

Most national governments, and, in fact, most people, now understand that we need to move away from polluting fossil fuels to renewable energy, even if they don't yet understand how our food and farming practices and our consumption habits are also major drivers of degeneration.

One major obstacle we all have to overcome, in both the industrialized and developing world, is the current lack of financial resources, investments, and political will to drive our great transition forward at the speed required. Another major obstacle is the lack of understanding in the Global North that our success depends upon helping our southern neighbors travel down the road with us to renewable energy and Regeneration. Of course, out-of-control political corruption, North and South, is a fundamental problem, misallocating, hoarding, and literally stealing essential resources. We all have to take back control over our personal lives and the economic and political institutions that shape our personal lives, or we will surely perish. We have no choice but to cut out the corrupt middlemen, businessmen, and bureaucrats and fundamentally reverse course—not in the far distant future, but now. We must make sure that the resources and the money—the billions (and eventually trillions) of dollars—that we need for this transition get into the hands of rank-and-file GND workers and enterprises and the Regenerators and Earth repairers, including small farmers, ranchers, and foresters, who will actually carry through the changes we need. But in order for this to happen, everyone needs to grasp that we will all rise or fall together. Runaway global warming will soon turn our planet into climate hell if we continue to allow corrupt, racist, and chauvinistic politicians to divide us, while their Big Oil, Big Pharma,

Big Food, Big Money, military-industrial complex backers hog the funds we need to make a just transition to a livable future. Without a massive increase in international awareness and solidarity and a systemic change in politics and economics, we will stall out on the road to Regeneration.

The as-of-yet unfulfilled "Green Fund" promises made at global climate summits to put $100 billion a year in the hands of developing nations to help them make the transition to a green economy must be seen for what they are: a just and long-overdue investment in our common survival and well-being. It doesn't matter if Europe, Japan, Scandinavia, or even the United States reaches its goal of 100 percent renewable energy in record time if India, China, Russia, and other less wealthy nations lag behind in transitioning from fossil fuel use. Greening the North in both the energy and farm sectors will not be enough to reverse global warming as long as greedy agribusiness corporations, multinational mining companies, and desperate farmers in Brazil, Indonesia, Africa, and the rest of the world continue cutting down rainforests and destroying carbon-sequestering wetlands and rangelands. It doesn't matter how quickly some nations adopt solar and wind power if the world's three billion more affluent consumers keep buying and consuming factory-farmed meat and animal products, along with grains, produce, and processed foods coming from industrial agriculture and food giants. It doesn't matter how quickly electric cars and trucks come on line if the world's industrialized farms and ranches continue plowing up and destroying topsoil, draining aquifers, spraying toxic chemicals, applying chemical fertilizers, and destroying the fertility and carbon-sequestering capacity of our forests and soils.

Starting Points

Can we realistically reduce greenhouse gas (GHG) emissions by 50 percent and scale up sequestration by fivefold within a decade, as part of a global Regeneration revolution? The good news is that we're already starting to do so.

On the renewable energy and energy conservation front, we must:

1. Continue to scale up the solar, wind, and renewable energy economy (which now provides more than 12 percent of global energy needs).

2. Increase vehicle fuel economy standards and replace our gas and diesel guzzlers with as many electric cars, buses, trucks, and trains as possible (transportation, including food transportation, now accounts for almost 14 percent of all global fossil fuel emissions).

3. Shut down all coal plants and completely eliminate coal use in the electricity and industrial/manufacturing sectors (coal currently is responsible for 30 percent of all global GHG emissions).

4. Aggressively move to green construction materials and practices and retrofit the world's buildings for greater energy efficiency (the global carbon footprint for buildings and construction amounts to 39 percent of all GHGs).

5. Build national and international "smart" electrical grids to efficiently utilize all of the renewable energy that the world will be producing.

Meanwhile regenerative land management practices (see chapter 4) are spreading across the globe. Building upon and scaling up these already existing, shovel-ready best practices of regenerative food, farming, reforestation, and land use on 10 to 20 percent or more of the world's croplands, rangelands, forests, and wetlands, while also repairing a significant proportion of our degraded and desertified wastelands, will buy us the time we need to convert to a 100 percent (or near 100 percent) renewable energy economy. This, in turn, will enable us to restore the essential carbon, methane, nitrogen, and hydrological cycles/balances between the atmosphere and our terrestrial and marine ecosystems—the major carbon and GHG sinks or repositories on the planet.

This Regeneration revolution will also help us, as we've seen, boost food quality, restore public health, eliminate rural poverty, and repair environmental destruction. In most cases we don't have to travel very far to find examples of the climate-friendly food, farming and land use practices that we need. There are living examples—generally small-scale, but in some cases larger-scale—of these best practices all around us, even if, up until now, most regenerative entities have received very little publicity. The kind of farms, ranches, gardens, forests, food hubs, wetlands, rangelands, and educational/training projects we need can be found in all 195 nations of the world, in every region, and in, or adjacent to, every one of the world's million cities, towns, and villages. Our job is to search out and map these centers of Regeneration, to support them, and to spread the word about them to the larger public. Then, with other Regenerators, we must build the local-to-international grassroots movement we need to scale them up. The road ahead is long and challenging, but objective conditions for a great Regeneration, facilitated by bold political action, as exemplified by the GND proposal in the United States, are now more favorable than they have been for decades.

Scaling Up Regenerative Best Practices

The most important thing that we—the global organic and Regeneration movement, including farmers, businesses, food-chain workers, and consumers alike—have to understand is that we're running out of time. We know the best way to do things, but our best practices are not spreading fast enough. We no longer have the luxury of remaining apolitical, of working primarily at the micro scale as we build up our organic permaculture farms, ranches, and local landscapes. We must move beyond wishful thinking, beyond believing that somehow we can convert the world's 570 million farms to organic and regenerative practices in the time frame we have left with just our own internal resources, expertise, and market clout. We need to move beyond ingrained habits and activist business-as-usual attitudes and join up with others, locally, nationally, and internationally. We need to turn every green-minded political activist into a conscious promoter of regenerative food and farming. And we need to turn every organic food consumer and farmer into a politically active proponent of a GND.

We need to scale up *best* practices, not just *marginally better* practices, on a critical mass of the world's four billion acres of croplands, eight billion acres of grazing lands, and ten billion acres of forests. To do this requires that we carry out a thorough and ongoing global mapping (i.e., locating and publicizing) of the best farm and land management practices that currently exist. The good news is that these best practices are potentially applicable to billions of acres, appropriate to different ecosystems, traditions, and farming conditions around the planet, and at the level sufficient to get us out of the predicament we face. In Eric Toensmeier's *Carbon Farming Solution*, in the anthology *Drawdown* (edited by Paul Hawken), and in practical farming journals across the world, such as those published by the Rodale Institute and Acres USA, there are numerous examples of farming, grazing, and agroforestry practices that are now sequestering and fixing from one to sixteen tons of carbon per acre *every year*. Given that we have twenty-two billion acres of croplands, rangelands, and forested areas to regenerate, along with several billion acres of wastelands, it is clear that we can reach our goal of sequestering five to twenty billion tons of carbon annually if we scale up best practices, especially agroforestry, silvopasture, and forest polycultures in the tropical and subtropical areas of the Global South, combined with regenerative organic agriculture and holistic livestock management in the Global North.

Articles in academic scientific journals that routinely underestimate the capacity for global soil carbon sequestration (saying, for example, that the best we can do is to mitigate 1 to 20 percent of current emissions, or a couple of billion tons of carbon per year) are ignoring the amazing sequestration practices already taking place on today's leading regenerative farms, ranches, and ecosystem restoration projects. At the same time, many academic "experts" are proclaiming, with little or no supporting evidence, that genetically engineered "climate-friendly" or drought-resistant crops or trees can significantly mitigate climate change. Other indentured academics praise marginally improved (corporate-sponsored) "climate-smart agriculture" practices, whereby commodity farmers use a reduced amount of soil-killing, biodiversity-destroying chemical fertilizers and pesticides, along with GMO seeds, $500,000 computerized tractors, and Big Data. In reality, none of these expensive high-tech schemes will eliminate food and farming emissions, much less rebuild natural soil fertility and supercharge soil sequestration.

The primarily low-tech, shovel-ready, affordable solutions that we need already exist in every nation and region. Millions of farms are already utilizing the traditional best practices of forest agriculture and forest gardens, organic and agroecological methods, holistic grazing, and soil conservation practices, augmented by recent innovations in permaculture, agroforestry, silvopasture, and landscape restoration. We don't need to invent new techniques. We simply need to identify, publicize, replicate, and scale up currently existing best practices utilizing farmer-to-farmer education and training, with major support and funding from the public and private sectors.

Our roadmap and planning must be based upon the understanding that our industrial and degenerative food and farming system is not just a small factor in global warming but a major cause of it (44 to 57 percent of all GHGs come from our globalized and industrialized food and farming system), and that traditional, organic, and regenerative practices—already being practiced in thousands of communities—offer not just slight mitigation but, in fact, a practical means to reverse climate change.

One of the unfortunate characteristics of most academic scientists today is their overreliance on the very limited number of on-the-ground farm studies published in peer-reviewed academic journals, rather than real-world developments, best practices, and breakthroughs. The same goes for mainstream journalists, who typically rely almost exclusively on these same academic experts and citations in the establishment press. The routine lack of imagination, overspecialization, and compartmentalization of science and journalism and the low estimates of the potential of regenerative carbon sequestration have now become especially dangerous, given our current climate emergency. Climate scientists, alternative energy advocates, politicians, and journalists who know little or nothing about farming, soil science, and carbon sequestration are not reliable sources of information on what is possible. Farmers, especially in the tropical and subtropical regions, sequestering five to fifteen tons of carbon per acre per year through agroforestry and silvopasture techniques, are typically not recognized in peer-reviewed scientific journals, nor are ranchers and others on the cutting edge of holistic management in Australia, North America, South America, and southern Africa, who have increased the soil carbon content on their farms and pastures sixfold or even twelvefold.

Breaking through the Final Roadblocks

We have the means and the working models we need for a global green economy and regenerative food, farming, and land use practices, but we lack the consumer awareness, political power, and economic resources to scale these practices up into a new Regenerative Commonwealth. Climate-denying fossil fuel companies, nation-states that depend upon petroleum exports, and the global industrial agriculture of the Big Food cartel, along with corrupt politicians they have in their pocket, are the major adversaries we must overcome. But false consciousness, cynicism, and fatalism are the other obstacles blocking our way to Regeneration. Parochial and nationalistic prejudices have spread like a cancer. Self-destructive lifestyles and deeply ingrained consumer habits and addictions have further sapped the energy of the global grassroots. The body politic and global civil society in most nations of the world are still divided, bogged down by pressing individual concerns and separated by cultural, economic, ethnic, and religious differences. Hypercompetitive trade relations and geopolitical conflict between the world's largest GHG emitters (the United States, China, Russia, India, and the European Union) drive politicians and corporations to delay or reduce investments in renewable energy and lower pollution and environmental standards so as to maintain corporate profits and the gross national product.

On the current (climate-destabilizing, land-degrading) food and farming front, the seemingly insatiable appetite of consumers in Europe, China, and United States for cheap (factory-farmed) meat and animal products drives the expansion of GMO animal-feed soybean plantations in Brazil, Paraguay, and Argentina, destroying the Amazon (often referred to as "the lungs of the planet"), as well as the once massive (and carbon-sequestering) grasslands and savannas of Latin America. Similarly, the Global North's lust for biofuels, cheap cooking oils for processed food, and cheap ingredients for body care products has spawned the destructive growth of palm oil plantations in Indonesia and many of the world's other tropical rainforest (and peatland) areas, undermining the rich biodiversity and massive natural carbon sequestration of these tropical forests.

Low-grade, subsidized food exports continue to be dumped by the Global North in the markets of the Global South, eroding public health, making it harder for traditional agroecological small farmers and graziers to maintain their regenerative or potentially regenerative farming practices, and undermining the South's ability to feed itself, employ its people, and maintain traditional social structures. That imbalance contributes to the ongoing forced migration of millions of people from the Global South. Fleeing violence and rural poverty (exacerbated by climate change), they are desperately attempting to cross seas and borders and find work in Europe, the United States, and other countries, setting off a wave of fear and xenophobia and enabling neo-fascist Earth destroyers and climate deniers like Trump in the United States and Bolsonaro in Brazil to gain power.

Yet perhaps the biggest obstacle we face on the road to Regeneration is the multitrillion-dollar military-industrial complex, in both the United States and other nations. Politicians complain that there's not enough money to finance climate change mitigation or a GND, but meanwhile there always seems to be money for massive military spending. Endless war and the trillion-dollar international trade in armaments are manifestations of not only the struggle to control fossil fuels, precious minerals, and other strategic resources but a self-destructive, nationalistic ideology of fear and distrust that justifies the status quo and pits countries, ethnic groups, and religious sects against one another. We can either continue with business as usual (my country and my people are better than your country and your people) and allow our multitrillion-dollar military-industrial complex to wage war and sow fear, or we can work together as part of a global New Deal, reverse global warming, and address all the other rapidly oncoming disasters of the modern age. Judgment Day has arrived. The time for action is now.

Given our shared predicament, my hope is that you're ready to step forward, take action, and join the growing ranks of Regenerators, either as an active leader in the front lines of advocacy and mobilization or as a supporter, donor, messenger, and practitioner of Regeneration in your local community. We are indeed cursed to be living in an age of crisis and Degeneration, but we are blessed to be living in an age of Regeneration as well. Never before have humans faced such a serious threat to our survival. But never before have we had the knowledge, the tools, the economic power, and the grassroots human resources to not only survive our terminal crisis, but to thrive. Never before have global subjective and objective conditions been so ripe for fundamental change. Revolutionaries and visionaries have been dreaming about global peace and transformation for two hundred years in the face of what turned out to be insurmountable odds. But now the balance of consciousness and power is shifting. We, the global civil society and the global body politic, finally have the opportunity to push forward with life-saving renewable and regenerative practices and policies as part of a world-changing coalition, both in the United States and globally. But the hour is late. What we do, or fail to do, today, tomorrow, and for the rest of this decade, is crucial. Our actions, not in a generation, but now and in the next ten years, will likely determine the future of our species.

Roadmap to Regeneration in the United States, 2020–2030

We have the outline of a plan. We need a mass mobilization of people and resources, something not unlike the U.S. involvement in World War II or the Apollo moon missions—but even bigger. We must transform our energy system, transportation, housing, agriculture and more.

STEPHANIE KELTON, ANDRES BERNAL, and
GREG CARLOCK, "We Can Pay for a Green New Deal"

The final months of 2018 will likely be remembered as the time when the United States and global grassroots finally began to awaken to the existential crisis posed by global warming. Part of this great awakening was no doubt due to the fact that violent weather, forest fires, drought, floods, water shortages, crop failures, and unusually prolonged heat and/or cold waves became the "new normal," striking home in both the Global North and the Global South, falling hardest on the poor and marginalized, but striking fear into the hearts of the middle and upper classes as well.

With international scientists finally dropping their customary caution and pointing out that the "end is near" in terms of irreversible climate change, the mass media, a significant number of global policy makers, and hundreds of millions of ordinary people seemed to simultaneously wake up across the world.

Young climate activists, under the banner of the Sunrise Movement in the United States and the Extinction Rebellion in the UK and other countries, sat in at politicians' offices and blocked streets and roadways, demanding immediate and bold action.[1] The Sunrise Movement captured headlines and mass public attention by calling for a sweeping change in

US federal policy: the Green New Deal. An international school strike, known as Fridays for the Future, initiated in Sweden by high school student Greta Thunberg, has begun to spread globally, with millions of students in over a hundred nations walking out of classes and organizing rallies and protests to demand bold action from their governments to reverse climate change.

But, of course, this great global awakening is just the beginning. As 350.org climate action leader Bill McKibben and others have pointed out, we now likely are at the point of our last chance to act on the climate crisis before it's too late. Here in the United States, we can't wait one or two more presidential election cycles before we take decisive action. Either we bring about bold economic and policy change, starting immediately, or we are doomed. Either we elect and rally behind insurgent green and social justice leaders and implement sweeping changes, or our global awakening in 2018 will be judged by future generations as too little, too late.

The Green New Deal under Attack

In the United States, the Sunrise Movement and Alexandria Ocasio-Cortez's Green New Deal (GND), though endorsed by more than a hundred members of Congress, as well as leading 2020 presidential candidates Bernie Sanders, Elizabeth Warren, Kamala Harris, and others, was immediately attacked as "too radical" or "utopian" by both climate-change-denying Republicans and neoliberals and indentured Democrats beholden to Big Oil and corporate agribusiness. In particular, the GND's proposition of achieving zero emissions by 2030 was dismissed as an impractical and dangerous measure that would wreck the economy and put millions of working-class people out of work.

If you read the GND proposal carefully, the criticism it has received is not justified, but it underlines the importance of being able to clearly explain to the American public and the global body politic exactly what we mean by a full-blown green energy and regenerative economy, with jobs for everyone willing to work and a just transition to *net zero emissions* by 2030. To gain and maintain majority support for policies such as the GND, we must be able to explain to everyday people not only the basics of reducing fossil fuel use and drawing down carbon through regenerative practices but also, as outlined earlier, how we can readily finance this great transition by increasing today's outrageously low taxes on the wealthy and large corporations and implementing a full menu of government appropriations, bonds, loan programs, jobs, and infrastructure projects, similar to the New Deal policies of the 1930s and '40s.

If we can properly explain what *net zero emissions* (as opposed to *zero emissions*) and a green economy with decent-paying jobs for all would mean, a critical mass of people and voters will likely see the GND for what it is: our last and best hope, a practical and comprehensive program based on sound science, public need, and commonsense survival.

Initial polls in the United States in December 2018 found that 81 percent of the public (Democrats, Republicans and Independents) basically supported the idea of a GND.[2] Later polls in 2019, even after prolonged criticisms (and misinformation) in the mass media, showed continuing majority support by 63 percent of Americans.[3] But, of course, the oligarchy and its indentured politicians and media spokespersons will continue to attack the GND. They will try to deny or ridicule the idea that we can actually change our current fossil fuel–dependent system, provide good jobs for everyone willing to work in reconstructing our urban and rural infrastructure and agriculture, and reverse climate change. To overcome these naysayers and gain critical mass, we will have to get organized and united as never before. We will have to carry out an unprecedented campaign of mass public education and mobilization, catalyzing a ballot box revolution that will put an end to the corporate domination of the US political system—and inspiring others around the world to do the same.

Zero and Net Zero Emissions

Unfortunately, most of the public, and even some of the early proponents of the GND, don't yet properly know how to explain what natural carbon sequestration actually means, what net zero fossil fuel emissions means, or what we're talking about when we say that regenerative food, farming, and land use, combined with renewable energy, can actually stop and then reverse, not just slow down, global warming.

In this regard, it is extremely important for Regeneration and GND advocates to be able to explain the difference between zero fossil fuel emissions and net zero fossil fuel emissions. *Net zero emissions* refers to the point in time at which we will be drawing down as much of our GHG emissions as we are still putting into the atmosphere and into our oceans. But *zero emissions*, in the minds of ordinary people, means literally just that—no fossil fuel or greenhouse gas emissions, period. A worthy goal to shoot for, but something that will likely take us more than ten years to achieve.

Net zero emissions takes into consideration the equivalent effect or impact of carbon drawdown. Of course we can't immediately, or even within a decade, move to global zero emissions by shutting down all cars, manufacturing, home heating and air conditioning, construction, and all commercial enterprises that utilize fossil fuels without wrecking the economy. But we can, even on the same tight ten-year time frame of 2020–2030, achieve net zero emissions through a combination of aggressive fossil fuel emissions reduction and aggressive regenerative carbon drawdown. Net zero GHG emissions will have the same practical impact on reducing global warming as zero emissions.

Of course, beyond net zero emissions, our long-term goal is to achieve *net negative* emissions, as soon as possible, whereby we begin to draw down and transfer 200 to 286 billion tons of excess atmospheric carbon—the dangerous legacy load of 820 billion tons of carbon from the atmosphere, where it's undermining climate stability—into our living soils and forests, where it will bring enormous benefits. Over a long period of time, this net negative process, as part of a new green economy, will enable our supersaturated oceans to release some of the excess carbon that they have absorbed from human-caused emissions, reducing the acidity of the oceans and restabilizing global habitat for marine life as well.[4]

Reaching Net Zero Emissions in the United States by 2030

As emphasized in chapter 2, don't let a bunch of numbers confuse you. Basically, what we have to do in the United States and the rest of the world over the next ten years is to cut fossil fuel emissions in half and then draw down the equivalent of the remaining GHG emissions into our soils, forests, and plants through regenerative practices.

In order to achieve the goal of net zero emissions in the United States by 2030, as called for in the GND, the most practical and achievable plan will be to reduce our current levels of net fossil fuel emissions from 5.7 billion tons of CO_2e to 2.75 billion tons of CO_2e, a reduction of 50 percent, while we simultaneously draw down and sequester in our soils and forests an equal amount (2.75 billion tons of CO_2e).[5]

In 2018, US GHG emissions amounted to approximately 16 percent of total global emissions (37.1 billion tons of CO_2e). In comparison, the US population of 330 million amounts to only 4.27 percent of the world's population. In other words, the United States is emitting approximately four times as much GHG per capita as the average person on the planet.[6] In fact, the United States is responsible for an estimated 28.8 percent of all human-derived global emissions since the onset of the industrial revolution in 1750.[7]

A GND for achieving a carbon-neutral economy in the United States by 2030 will necessarily involve eliminating 45 to 60 percent of our current 5.75 billion tons of CO_2e fossil fuel emissions, while sequestering the remaining two to three billion tons of CO_2e through regenerative agriculture, reforestation, and ecosystem restoration. This represents an ambitious but realistic goal, according to numerous experts and current best practices, assuming we can generate sufficient political pressure to force the White House, Congress, and state and local governments to reject business as usual and take bold action.[8]

Categories of Emissions

The EPA breaks down the sources of the United States's gross GHG emissions into five broad categories: transportation (29 percent), electricity production (28 percent), industry (22 percent), commercial and residential (12 percent), and agriculture (9 percent).[9] Taking the agriculture category at face value, you'd likely infer that food, farming, and land use are no more than a minor factor (9 percent) in the United States' contribution to the global climate crisis. However, if you look more closely at the carbon or CO_2e "footprint" of the food, farming, and land use sector as a whole (including fossil fuels used in on-farm production, food and crop transportation, food processing, packaging, and refrigeration, as well as the chemical inputs of pesticides and chemical fertilizers, off-gassing of CO_2, methane, and nitrous oxide from soils and landfills, and destruction of wetlands and soil organic carbon), you start to realize that food, farming, and land use are actually responsible for almost half of all US GHG emissions, not just the 9 percent attributed by the EPA and the USDA to "agriculture."[10]

On the other hand, organic and regenerative farming and land management practices and forest growth in the United States are routinely overlooked as being important solutions to global warming and climate change. Properly managed lands and forest growth actually draw down a considerable amount of excess CO_2 from the atmosphere. Currently they sequester 714 million tons of CO_2e (or 11 percent of US gross emissions) annually, even according to the EPA, and even in their currently degraded condition.

We Need Net Zero Emissions by 2030, Not 2050

There is a debate in progressive political circles about whether we should adopt a more conservative goal, as put forth by the Intergovernmental Panel on Climate Change (IPCC) and most nations, to achieve net 45 percent reductions in GHGs by 2030 and net zero emissions by 2050, or whether we should instead aim for a much more ambitious goal, in line with the goals of the GND, to achieve net zero emissions by 2030.

A number of nations have already pledged to reach net zero emissions before 2050, including Bhutan (which has already achieved net zero emissions), Norway (2030), Uruguay (2030), Finland (2035), Iceland (2040), and Sweden (2045). The state of California, too, can be added to this list (2045). The European Union is currently operating under a net zero 2050 timeline but will likely set a stricter goal soon.[11]

Let's now look in more detail at how we can reduce fossil fuel emissions in the United States by 45 to 60 percent in the next decade through energy conservation and making the transition to renewable energy. Following that, let's look at how we can draw down or sequester the remaining two to three billion tons of GHGs that we will still be emitting in a decade, so as to achieve net zero emissions.

US Roadmap Part One:
Reducing Fossil Fuel Emissions by 45 to 60 Percent

The United States is fortunate to have the natural resources to help lead the global community in a transition to a green energy future, complemented by regenerative agriculture and land use. We not only have some of the best wind, solar, geothermal, hydro, and biomass resources on Earth, but our forests, soils, farmlands, grasslands, wetlands, and marine ecosystems have the inherent capacity, if properly managed and regenerated, to sequester as much CO_2e as we are currently emitting, and even more. Perhaps most important of all, we have a new generation of youth, personified by the Sunrise Movement, supported by a new wave of climate-conscious, insurgent politicians, such as Bernie Sanders and Alexandria Ocasio-Cortez, ready and willing to take the lead.

On the renewable energy front, the United States, under a new administration in 2021 and beyond, will need to step up the pace. We must rapidly expand the US solar, wind, and renewable energy economy, which in 2017, according to the EPA, provided approximately 13 percent of our energy needs, including 22 percent of our electricity. As we ramp up renewables, we must phase out coal, oil, gas, and nuclear power plants, as rapidly as possible. Germany, with a powerful economy similar to that of the United States, has been operating under a plan for ten years to reduce emissions by 55 percent by 2030, and will likely soon be raising its goals for emissions reductions even higher. If the United States sets a goal for a 60 percent reduction in fossil fuel use/GHG emissions by 2030, similar to that of Germany, we should be able to sequester the remaining 40 percent of GHGs through regenerative food, farming, reforestation, and ecosystem restoration practices, enabling us to reach net zero emissions (carbon neutrality) by 2030. Although 60 percent in emissions reductions is an achievable goal, as we will demonstrate below, even with 45 to 50 percent reductions we can still reach carbon neutrality by maximizing regenerative agriculture and forest/land management practices.

To reduce fossil fuel use and GHG emissions by 45 to 60 percent over the next decade, we will need to generate 75 to 85 percent or more of our electricity (which now releases 28 percent of our total emissions) with renewables, basically shutting down coal generation for electricity. This will reduce current overall emissions by approximately 20 to 24 percent. Energy conservation measures across all sectors (utilities, transportation, buildings, manufacturing, agriculture) will need to go along with this renewable energy revolution in the electricity sector.

According to a comprehensive study published in 2015 by Mark Z. Jacobson and a team of experts in the peer-reviewed journal *Energy and Environmental Science*, all fifty states in the nation have the potential to convert their fossil fuel–based electricity, transportation, heating and cooling, and industry systems to ones powered entirely by wind, water, and sunlight, replacing 80 to 85 percent of existing fossil fuel and nuclear energy by 2030 and 100 percent by 2050. In terms of the economic impact of this mass conversion to renewable energy on employment, the study's authors state: "Over all 50 states, converting would provide ~3.9 million 40-year construction jobs and ~2.0 million 40-year operation jobs for the energy facilities alone, the sum of which would outweigh the ~3.9 million jobs lost in the conventional energy sector."[12]

In terms of technological innovation, according to numerous studies, it is now cheaper and more profitable to build and operate electricity generation systems using solar and wind power than it is using coal, nuclear, or petroleum power.[13]

But in order to replace coal, natural gas, nuclear, and petroleum as our primary power sources, our national (and international) electrical grid infrastructure will have to be rebuilt to facilitate decentralized power production and electricity sharing across regions. Also, we will obviously need to stop building more fossil fuel infrastructure (including pipelines), curtail oil and natural gas exploration and extraction, phase out polluting power plants, and electrify manufacturing, transportation, and heating. All of these measures mean leaving most, and eventually all, remaining fossil fuel reserves in the ground.

To pay for this transition, we will need to transfer massive government subsidies from fossil fuels to renewables and, at the same time, ensure a just transition and program of retraining for four million current workers in the fossil fuel sector, as outlined in the GND. If we don't ensure a just transition (job retraining, job replacement, and/or retirement) for fossil fuel workers, we will likely never gain the political support for the GND that we need.

In the transportation sector (29 percent of current emissions), we will need to double or triple vehicle fuel economy standards and replace our gas and diesel guzzlers with as many electric cars, buses, trucks, tractors, and trains as possible so as to achieve 50 percent market share for electric vehicles by 2030. In order to do this, we will need to pay consumers, businesses, and municipalities a subsidy to switch over to electric vehicles and electrified mass transportation. This could potentially cut overall emissions by approximately 50 percent in the transportation sector. According to *MIT Technology Review*, given battery technology advances and cost reductions (electric cars will soon be cheaper to buy and operate than gas-driven vehicles), over half of new auto sales in 2040 will be electric vehicles.[14]

Beyond automobiles, a growing number of nations are leading the way in terms of converting petroleum-driven buses, trucks, and trains to electricity. According to a report released at the San Francisco Global Climate Action Summit in 2018:

> Every 5 weeks, China adds a fleet of electric buses equivalent to the entire London bus fleet—9500 buses. Technologies are now market ready, societally acceptable and economically attractive to reduce greenhouse gas emissions from transport by 51% by 2030, through electric vehicles, mass transit and adapting the global shipping fleet. . . . However, the transformation will slow dramatically without strong national and city policies, for example setting target dates to ban internal combustion engines.[15]

In the industrial and manufacturing sector, including heavy industry, light industry, feedstocks, and food processing (22 percent of all fossil fuel emissions), we will need to reduce coal and petroleum use by at least 50 percent through dramatic increases in recycling rates, switching over as quickly as possible to electrical power generated by renewables, and efficiency improvements, such as "making products more material-efficient . . . extending lifespan and reducing weight."[16] In the light industry sector, including food, textile, wood, printing, and consumer products manufacturing, as well as more fossil fuel–intensive industries such as steel, aluminum, cement, and plastic production, according to experts, we can reduce overall emissions by 50 percent using current technologies and efficiencies.[17] Of course, addressing overconsumption and waste on the part of consumers, especially more affluent consumers, will need to be part of this mission.

A transition from cement to timber in the construction industry (a growing number of buildings, even high-rise buildings, are now being built from wood, using new techniques) could eliminate 5 to 6 percent of all GHG emissions. Similar climate-friendly changes in the manufacturing, construction, and industrial sector will reduce emissions by another 10 percent, for a cumulative total reduction of emissions of 45 to 49 percent in the electricity/utilities, transportation, and manufacturing sectors.

Residential and commercial buildings now account for almost 11 percent of all fossil fuel use in the United States. We can achieve a 50 percent reduction in emissions in this sector with retrofitting, insulating, changes in building codes, and greater energy efficiency, utilizing heat pumps, solar power, heat storage, and district heating systems based upon renewable energy. This retrofitting of our buildings will reduce our overall emissions by another 5 to 6 percent, while creating millions and millions of new jobs.

This brings us to 50 to 55 percent in overall fossil fuel emissions reductions by 2030. We can achieve a further 5 to 10 percent overall emissions reduction in the food and farming sector by reducing the consumption of fossil fuel–derived materials and products (plastics, food packaging, highly processed foods), by eliminating food waste and clothing waste, by recycling organic waste instead of dumping it into landfills, where it releases methane and CO_2, and by drastically reducing methane and nitrous oxide emissions from fracking, natural gas, and chemical-intensive agriculture and factory farm inputs and practices (diesel fuel, chemical fertilizers, and petroleum-derived pesticides). Sixteen percent of all US GHG emissions comprise the potent heat-trapping gases nitrous oxide and methane—a significant percentage of which come from chemical-intensive industrial agriculture emissions from heavy pesticide and fertilizer use.[18]

Implementing all of these transformations/reductions in the electricity, transportation, manufacturing, residential and commercial buildings, food, farming, and consumption sectors, as called for by the GND, can enable the United States to basically match the emissions reduction goals of Germany, with a 60 percent decrease in GHG emissions by 2030 (from 5.9 billion tons of CO_2e to 2.36 billion tons).

The remaining 40 percent of net reductions (2.36 billion tons of CO_2e) required to reach net zero emissions and a carbon-neutral USA by 2030 will need to be sequestered in our 1.9 billion acres of croplands, pasturelands, rangelands, wetlands, forests, urban landscapes, and vegetation through regenerative food, farming, forestry, land management, and ecosystem restoration practices. Let's now look in more detail at a Regeneration and carbon sequestration plan for the next decade.

US Roadmap Part Two: Sequestering Remaining CO_2e through Regenerative Food, Farming, and Land Use Practices

Utilizing satellites, surveys, and other sources, the USDA categorizes the 1.9 billion acres of the Lower 48 (i.e., all the states excluding Alaska and Hawaii) as follows: 654 million acres of pasture or rangeland (much of which is degraded), 539 million acres of forest (much of which needs to be reforested), 392 million acres of cropland (most of which is degraded in terms of soil carbon), 169 million acres of "special use" lands (parks and national/state forests), 69 million acres of urban land, and 69 million acres of "miscellaneous."[19]

Currently, as estimated by the EPA, the Lower 48 (1.9 billion acres) are sequestering 714 million tons of CO_2e (or 11 percent of US gross GHG emissions). To reach our goal of net zero emissions by 2030 (assuming energy conservation and renewable energy can reduce gross emissions by 60 percent), these 1.9 billion acres, or rather a significant percentage of these acres, will have to be regenerated and reforested over the next decade so that they can sequester approximately three to four times as much atmospheric carbon as they are currently sequestering. Looking at scaling up existing best practices, we can see that this great sequestration and recarbonization of our soils and biota is indeed possible.

Let's look at the practices (and the math) of potential carbon sequestration (and reduction of methane and nitrous oxide emissions) on the 1.9 billion acres of US farmland, pastures, rangelands, forests, and other landscapes by 2030.

Regenerating US Pasture and Rangeland

US pasture and rangeland (654 million acres) covers more than one-third of the Lower 48. One-quarter (158 million acres) of this acreage is administered by the US government and is usually open to livestock grazing by ranchers for a fee. Another 127 million acres that the EPA classifies as croplands are used by farmers to grow animal feed for livestock. This means that the livestock and livestock feed portions of our agricultural lands adds up to 781 million acres, 41 percent of all the land in the Lower 48.

The majority of these pastures and rangelands were once a diverse landscape—grasslands and natural prairie covered with native (deep-rooted) grasses, trees, bushes, and plants. This carbon-rich, climate-friendly landscape sequestered large amounts of atmospheric carbon, supported biodiversity and wildlife, and efficiently infiltrated rainfall and snowmelt into the topsoil and groundwater, springs, and aquifers. Before the advent of the plow and the repeating rifle and the ruthless occupation of Native lands, large herds of migratory buffalo, elk, deer, and other mammals grazed on the grasslands as they moved across the continent, while millions of "keystone species," including beavers (wetlands builders), prairie dogs (soil excavators), and wolves (forcing grazing herds to band together and killing off sick and diseased animals), worked in natural harmony to keep the landscape regenerated and hydrated.

At the present time, most of these 781 million acres have been plowed under, deforested, and/or overgrazed, leaving them eroded, degraded, and lacking in terms of soil organic carbon, soil fertility, and biodiversity. America's once healthy pasturelands and farmlands have become major greenhouse gas emitters, rather than soil carbon sinks or repositories. But with regenerative changes in grazing practices and livestock management, including switching cattle and herbivores away from chemical-intensive, fossil fuel–intensive GMO grains to a 100 percent grass-fed diet, and moving poultry and pork from confinement to free-range pasture, raised on a diet of organic and regeneratively produced grains, we can bring US rangelands and pasturelands back to full life and vitality.

Cattle and other herbivores such as sheep, goats, and buffalo should be outdoors, grazing on pasture grass, while omnivores such as poultry and pigs should be moved out of confinement and raised outdoors in a free-range or agroforestry setting, getting some of their nutrition/food in their pastures or wooded paddocks, while getting most of their nutrition from grains and forage that have been grown in a regenerative manner (no-till, cover-cropped, alley-cropped, biodiverse, chemical-free, agroforestry). Cattle and other herbivores will thrive and produce healthier meat and dairy products once they return to a 100 percent grass diet, as will chickens and pigs raised in a natural free-range environment. And consumers, once they understand the nutritional, environmental, climate, and animal welfare superiority of grass-fed and pastured meat and dairy, will increasingly choose to buy these products, especially if current subsidies—direct and indirect—for factory farms and factory-farmed animal feeds are eliminated.

Of course, all of this will require major subsidies for farmers and ranchers (including guaranteed fair prices, supply management, and payments for soil conservation) as well as changes in consumer purchasing and consumption, including a drastic reduction in the purchasing and consumption of factory-farmed meat, dairy, and other grain-fed animal products (chicken, poultry, and factory-farmed fish).

To carry out this restoration on a large enough scale, we will have to put an end to wasting millions of acres of our valuable farmlands to grow grain for herbivores (cows, sheep, goats, and bison)—animals that should not be eating grains at all. We will also need to stop sacrificing thirty-eight million acres of our valuable farmlands to the production of ethanol and biodiesel from corn and soybeans and instead convert these row-crop commodity farms back into diverse crop production and grazing. The process of producing ethanol and biodiesel from GMO corn and soybeans, contrary to industry claims, actually uses up more fossil fuels in its growing and production cycle than it saves by allowing us to burn ethanol or biodiesel in our cars.

Regenerative management of these pasturelands and rangelands will utilize soil-building techniques such as no-till farming, multispecies cover cropping, roller crimping (breaking the plant stalks and leaving them on the field rather than plowing or spraying pesticides when the cover crops mature), and grazing animals holistically and rotationally.[20] Once restored and under regenerative management, these lands can sequester approximately twelve tons of CO_2e per acre per year.

Guaranteed subsidies for soil conservation practices, a waiver of grazing fees on properly grazed federal lands, and fair prices (coupled with supply management) for farmers and ranchers for their meat, dairy, and grains are some of the key policies we will need to implement after the 2020 elections in order to promote regenerative, carbon-sequestering management of the majority of these 781 acres of pasturelands, rangelands, and animal feed croplands. The federal farm and soil conservation policies that we will need to fund in order to achieve a carbon-neutral economy by 2030 include the following:

Expansion of the Conservation Stewardship Program and the Environmental Quality Incentives Program, with billions of additional dollars a year to increase regenerative practices such as cover cropping, prescribed grazing, riparian buffers, and no-till farming.

Expansion of the Conservation Reserve Program (CRP) to include 100 million acres by 2030, raising rental payments made to farmers, and promoting regeneration practices, including agroforestry and holistic grazing, on these CRP lands.

Expansion of the Regional Conservation Partnership Program to substantially increase the acreage that farmers place into agriculture conservation and wetlands easements.

A major increase in the funding for research into conservation and holistic grazing, focusing on research into the reduction of carbon emissions in the agricultural sector and eliminating degenerative factory-farm production methods, as well as research dedicated to soil health.

Billions of dollars in increased incentives for local and regional food systems, as well as incentives for reforestation, regenerative forest management, and restoration of coastal wetlands. We will need to reforest over 65 million acres by 2030, on a combination of Forest Service, Bureau of Indian Affairs, and other federal lands, as well as on state, local, tribal, and nonprofit-owned lands. By 2050 we will need to reforest more than 250 million acres.

We must protect millions of at-risk acres of federal, state, local, tribal, and other lands by 2030 using forest management, controlled burns, and holistic grazing practices to reduce the risk of catastrophic wildfires and to increase forest health/resilience. We need to plant an average of fifty million trees per year in urban areas across America to reduce the heat island effect and protect communities from extreme weather. In addition, we need to invest in wood product innovation and in biochar, creating jobs in rural and urban communities. Besides these measures, we need to restore or prevent the loss of 12 to 25 million acres of coastal and inland wetlands by 2030.

If holistic grazing and livestock/pasture management best practices were carried out on just a quarter of total pastureland, rangeland, and animal feed cropland in the United States, we would still be able to sequester 2.34 billion tons of CO_2e —approximately 100 percent of the carbon sequestration we need (in combination with a transition to renewable energy) to reach net zero emissions by 2030.

Regenerating US Cropland

US cropland (392 million acres) includes 52 million acres idled or lying fallow at any given time, 38 million acres used for corn ethanol or soy biodiesel, 77 million acres for human food for US consumers, 127 million acres for livestock food crops (especially corn and soy), 22 million acres for wheat exports, 14 million acres for cotton (fiber and animal feed), and 69 million acres for other grains and food exports. Yet despite its enormous agricultural production, the United States imported 15 percent of its food and beverages in 2016, including 30 percent of its fruits and vegetables.

Disregarding the 127 million acres of cropland used for livestock grains and fodder, which we have discussed in conjunction with pasture and rangeland above, the United States' 265 million acres of additional cropland can potentially be regenerated in order to store more carbon and improve fertility, water quality, biodiversity, food safety, and food quality or nutrition.

Traditional organic crop farming (no chemicals, cover cropping, minimum or no tillage, use of natural fertilizers) can sequester CO_2e at a rate of up to 5.7 tons of CO_2e per acre per year.[21] However, Dr. David Johnson's New Mexico lab and field research on regenerative compost shows that high-fungal-content, biologically rich, semi-anaerobic compost and compost extracts produce not just very high crop yields but also massive carbon sequestration, with rates of over four tons of carbon (fifteen tons of CO_2e) per acre per year. As Dr. Johnson notes, if these compost practices were scaled up on the world's four billion acres of croplands, "the entire world's carbon output from 2016 could be stored on just 22 percent of the globe's arable land." Perhaps not coincidentally, Johnson's methods mirror traditional and indigenous compost and agroecological farming practices utilized in India and other regions.[22]

If traditional organic crop practices were implemented on all of the 265 million acres of US cropland (again, not counting land given over to animal feed crops), we could sequester 1.3 billion tons of GHGs. If organic practices were employed on just 50 percent of these croplands, we could sequester 650 million tons. With traditional organic practices on just one-quarter of this cropland, we could sequester 325 million tons.

But if advanced organic practices like Dr. Johnson's were implemented, we could sequester 3.9 billion tons a year on 265 million acres, or 1.95 billion tons on half of this acreage, or almost 1 billion tons of GHGs on one-quarter of this acreage.

As a conservative estimate, with a combination of traditional organic and advanced organic methods on one-quarter of US cropland, we will be able to achieve 663 million tons of CO_2e sequestration—approximately one-quarter of what we need.

Necessary measures to transform US crop production will include increasing the market share of organic food from its current 5.5 percent of all food sales and 10 percent of all produce (fruit and vegetable) sales to 50 percent of all sales by 2030. At the same time, we will need to convert thirty-eight million acres of corn (ethanol) and soybean (biodiesel) crops back into multispecies perennial grasslands and pasture and/or organic multispecies grain production. We will also need to implement soil restoration, regeneration, and agroforestry practices on our fifty-two million acres of idle or fallow land, utilizing government programs to subsidize farmers for restorative and regenerative practices.

Regenerating US Forestlands

US forestland (539 million acres), or rather "unprotected" forests and timberlands in the terminology of the USDA, account for one-quarter of the land in the Lower 48. These 539 million acres do not include the "special use" protected or semi-protected forest acreage in national parks (29 million acres of land), state parks (15 million acres), or wilderness and wildlife areas (64 million acres), or the "miscellaneous" ("low economic value") acres of trees and shrubs located in marshes, deserts, and wetlands. Nor does this acreage include trees in urban areas.

If we count all these other forested (or "treed") areas, however, forests comprise one-third of the total US land area. That may seem like a lot, but keep in mind that forests covered half the country prior to European settlement.[23]

The EPA estimates that US forests currently sequester approximately 9 percent of all US GHG emissions (531 million tons of CO_2e) every year. Over the next ten years, in order to reach carbon neutrality, we will need to embark upon a major program of reforestation and afforestation—preserving, expanding, and improving our forests (both private and publicly owned) and tree cover (both urban and rural).

According to the rather conservative projections made by the Nature Conservancy, reforestation of forty to fifty million acres in the United States could reach three hundred million tons of additional CO_2e captured per year by 2025.[24] But according to a more recent study by Dr. Thomas Crowther and others, mentioned in chapter 4, the United States has 254 million acres of degraded forests or treeless landscapes (excluding croplands and urban areas) that could be reforested, especially in the South, Southeast, and Northeast regions of the country. These 254 million reforested acres could potentially sequester, using the Nature Conservancy projections, 1.5 billion tons of GHGs annually.[25]

Even if we reforest only one-quarter of the potential area that could be reforested in the United States by 2030, we will still be able to sequester 375 million tons of CO_2e—approximately 15 percent of what we need.

As a recent article titled "Let's Reforest America to Act on Climate" points out: "Under the original New Deal, the Civilian Conservation Corps planted three billion trees and employed three million workers in the process. America is well positioned to advance a similar effort again, with almost 20 million acres of recently disturbed land needing reforestation."[26]

"Special use" lands (169 million acres), including parks, wildlife areas, highways, railroads, and military bases, include millions of additional acres suitable for reforestation and afforestation, as identified by Crowther and others.

Urban areas (69 million acres) make up 3.6 percent of the land area of the Lower 48 but include 81 percent of the population (19 percent of people live in rural areas). Urban areas are growing by a million acres a year. Lawn areas in US cities and towns are estimated to include forty million acres of turf grass, covering 1.9 percent of the land. Although Crowther and others do not include urban areas in their totals for land that could be reforested, obviously millions of acres in urban areas are suitable for planting trees, which would then sequester carbon, reduce summertime urban temperatures, and provide shade, food, and habitat for humans, pollinators, and animals. In the United States, we should set a goal for planting 500 million new trees in urban areas by 2030.

Regenerating So-Called "Miscellaneous Lands"

"Miscellaneous lands" (69 million acres) are categorized by the USDA as having "low economic value." These lands include cemeteries, golf courses, and airports, but also marshes and coastal wetlands. Contrary to the USDA's assessment, the nation's marshes and wetlands are enormously important in terms of sequestering carbon, filtering pollution, buffering hurricanes, preserving water quality, and providing habitat for fish and wildlife. As part of a national campaign of ecosystem restoration and carbon sequestration in the United States, we will need to restore millions of acres of wetlands, marshes, and marine ecosystems. It is estimated that the continental United States (not including Alaska) once had 220 million acres of wetlands, most of which have now been drained or destroyed. Restoring 12 to 25 million acres of marshlands and wetlands in the Lower 48 would sequester 75 to 150 million tons of CO_2e annually.[27]

The Bottom Line for US Carbon Neutrality

The bottom line for achieving carbon neutrality in the United States by 2030 is to basically reduce fossil fuel emissions by 45 to 60 percent in our electricity, transportation, housing, construction, and manufacturing sectors, in line with what other advanced industrial nations such as Germany are undertaking, while simultaneously carrying out the regenerative, carbon-sequestering agriculture and land use practices outlined above. With changes in livestock and pasture management on just a quarter of total pastureland, rangeland, and animal feed cropland (781 million acres), by 2030 we can sequester more than 2.34 billion tons of CO_2e annually. With changes in management, utilizing organic and advanced organic methods, on a quarter of our 265 million acres of croplands (not counting land used to produce animal feed), we can achieve an additional 663 million tons of CO_2e sequestration. With reforestation and afforestation on 25 percent of the 254 million acres of degraded forests or treeless landscapes (excluding croplands and urban areas) in the United States, we can sequester an additional 375 million tons of CO_2e. Restoration of wetlands can sequester an additional 75 to 150 million tons. Altogether, by 2030, this great regeneration will sequester 3.4 billion tons of CO_2e annually, enough to enable the United States to reach carbon neutrality, even if the country only manages to reach 45 percent in fossil fuel reductions, rather than the 60 percent that Germany and a number of other nations will achieve.

Altogether, with the ongoing restoration and regeneration of our 1.9 billion acres of pasturelands, rangelands, croplands, forests, and wetlands—driven by changes in public policy, consumer demand, and farmer/land management innovation—we, as part of a GND, can lead the United States (and, by example, the world) away from climate catastrophe to carbon neutrality. This will then prepare us to keep moving forward beyond 2030: to draw down enough excess carbon from the atmosphere into our revitalized soils, forests, and plants to reverse global warming and restore our precious environment and climate. But the hour is late. We need a GND and a Regeneration revolution. And we need to step up our public education, coalition building, direct action, and electoral insurgency now.

Political Power Now: Greening the White House and the Congress

We have no choice but to move boldly forward with a system-changing GND in the United States and other nations, infused with the goal of 100 percent renewable energy and a massive scaling up of regenerative food, farming, and land use policies and practices. But if we hope to gain the support we need from working people and lower-income communities, renewable energy and regenerative food and farming must be delivered as part of a popular overall package for a *just* transition that includes full employment, livable wages, universal health care, debt relief, and free public education as well.

Like it or not, what the United States does or does not do in the 2020 election (and the 2020–2030 decade) is crucially important. We need a new president, we need a new green-minded majority in the House and the Senate, and we need new green and Regeneration-supportive government officials and public policies in all of our states, counties, cities, and towns. This means that our number one priority, given our limited timeline, must be to join and help build a mass movement to take power in Congress and the White House in 2020 and 2022.

Fortunately, we already have the initial public support (63 percent of people in the United States currently support the GND), grassroots leaders (the Sunrise Movement and a growing activist rainbow of movements and Regenerators), and a new insurgent group of political leaders who share our vision, who will be welcomed by an already Regeneration-minded movement and government in the nations around the world. We already have 90 or more of the 435 members of the House of Representatives who have endorsed the GND, along with a dozen high-profile senators. All of the leading Democratic Party candidates for president in 2020, including Bernie Sanders and Elizabeth Warren, have endorsed the GND. For the first time ever, climate change has become a major electoral theme in the United States and other nations.

Jump-Starting the Green Machine

The next step prior to the crucial November 2020 presidential and congressional elections in the United States (and elections in other nations) is to build mass awareness at the local, state, and congressional levels. We need local GND/Regeneration committees; we need speaker's bureaus; we need media teams; we need fundraisers, coalition builders, and grassroots and grasstops lobbyists; and we need online and, most important, on-the-ground activism (petitions, teach-ins, door knocking, protests, electoral campaigns, ballot initiatives). We must start now to build broad-based, powerful, bipartisan if possible, statewide and national coalitions for a GND that highlight not only renewable energy, but regenerative food, farming, and land use policies and practices as well.

We have no choice but to break down the issue silos that divide us—we don't have time for anyone to think, "My issue is more important than your issue," or "My constituency is more important than your constituency." We must connect the dots, create synergy, and unite a critical mass of heretofore single-issue, limited-constituency movements (climate, peace, labor, health, environment, food, farming, and social justice). At the same time, on the political front we must strive to bring together for discussion and common action progressive Democrats and conservation-minded Independents, Republicans, and Libertarians. We must build awareness and cooperation in a survival-oriented united front that can elect green and Regeneration-minded majorities in both urban and rural districts. Breaking down walls and issue silos, we must convince renewable energy and progressive political activists that regenerative food, farming, and land use practices and policies are essential, while at the same time getting food, farming, and environmental/conservation activists to understand that we must all become climate activists and renewable energy advocates and we must all get involved in political action.

The Power of One in Catastrophic Times

No doubt you've heard something like the message of this book before. I've personally been writing and campaigning around a host of life-or-death political, food, farming, and environmental issues like these for fifty years, starting with the threat of nuclear annihilation in the Cold War, the civil rights movement, and the Vietnam War in the 1960s. The exciting, world-changing difference now is that objective conditions are finally ripe for a Regeneration revolution in the United States and around the world. What I've said and written before about the environment, food, health, politics, war, and peace, with every ounce of knowledge and passion I could muster, was basically true. It's just that we, the global grassroots—farmers and consumers, students and workers, and our political and activist leaders—weren't quite ready yet. The crisis of the past fifty years hadn't yet reached its present intensity. In addition, up until now, we didn't have a workable plan, strategy, and tactics. We didn't have a GND or a set of radical political leaders at the federal level to rally behind. We didn't have grassroots leaders in every community like those that we have now. We didn't have a full understanding of the relationship between food, farming, land use, soil health, fossil fuels, climate change, deteriorating public health, environmental degradation, justice, international relations, war, and peace. Now we do. Now we can connect the dots and move forward together, not just in one region or country, but globally.

Here's an excerpt from a speech I gave twenty-five years ago, on September 24, 1995, at the US-Canada International Joint Commission on the Great Lakes. I think my message from then is even more relevant today:

> The time bomb we call the future is ticking away even as we consider these matters. We have no time to lose. The time for standing around and feeling inadequate or frustrated is over. If you've been waiting for new movement leadership and new ideas to arrive,

wait no longer. Look in the mirror, look at the people around you today. Go back to your community and form an affinity group of like-minded individuals, people whom you feel good about. Work with people who will make your social change efforts effective as well as fulfilling, and yes, even joyful. People bold enough to take on the corporate Global Lords, yet humble and grounded enough to practice what they preach. Once properly grounded, link up your core group and your outreach and coalition-building efforts with other compatible groups in your community, county, state, and region. If you're not exactly certain of how to go about getting organized in your community, then search out the activist "coaches" and social-change movement "veterans" who are willing to help you. Don't mourn about the state of the world or the state of your individual soul! Organize! There's only one reason for joining up in the worldwide movement to save the planet and build a more democratic and ethically sound commonwealth: because it's the best way to live.

It makes a great deal of difference what you and I do as individuals in our everyday lives. It makes a difference how you and I behave in the marketplace, and in the realm of civil society and politics. How we act, what we talk about with family, friends, neighbors, and coworkers. How we spend our money and our precious spare time. How we raise our children. What we read and share and write as we sit in front of our computers and cell phone screens. Which groups we join, support, and donate money to. Which politicians we lobby and vote for.

Never underestimate the power of one individual—yourself. But please understand, at the same time, that what we do as individuals will never be enough. We have to get organized, and we have to help others, in our region, in our nation, and everywhere, build a mighty green Regeneration movement. The time to begin is now.

RESOURCES

Key Websites

Regeneration International

    Website: regenerationinternational.org

Organic Consumers Association

    Website: organicconsumers.org

Sunrise Movement

    Website: sunrisemovement.org

Recommended Reading

Biosequestration and Ecological Diversity: Mitigating and Adapting to Climate Change and Environmental Degradation, by Wayne A. White

Burn: Using Fire to Cool the Earth, by Albert Bates and Kathleen Draper

Call of the Reed Warbler: A New Agriculture, A New Earth, by Charles Massy

The Carbon Farming Solution: A Global Toolkit of Perennial Crops and Regenerative Agriculture Practices for Climate Change Mitigation and Food Security, by Eric Toensmeier

Compendium of Scientific and Practical Findings Supporting Eco-Restoration to Address Global Warming, a multivolume work compiled by Biodiversity for a Livable Climate

Cows Save the Planet: And Other Improbable Ways of Restoring Soil to Heal the Earth, by Judith Schwartz

Defending Beef: The Case for Sustainable Meat Production, by Nicolette Hahn Niman

Dirt to Soil: One Family's Journey into Regenerative Agriculture, by Gabe Brown

Drawdown: The Most Comprehensive Plan Ever Proposed to Reverse Global Warming, edited by Paul Hawken

Geotherapy: Innovative Methods of Soil Fertility Restoration, Carbon Sequestration, and Reversing CO_2 Increase, edited by Thomas J. Goreau, Ronal W. Larson, and Joanna Campe

Grass, Soil, Hope: A Journey through Carbon Country, by Courtney White

The Great Climate Robbery: How the Food System Drives Climate Change and What We Can Do about It, by GRAIN

Healing Earth: An Ecologist's Journey of Innovation and Environmental Stewardship, by John Todd

The Hidden Half of Nature: The Microbial Roots of Life and Health, by David R. Montgomery and Anne Biklé

Poisoning Our Children: The Parent's Guide to the Myths of Safe Pesticides, by André Leu

Restoration Agriculture: Real-World Permaculture for Farmers, by Mark Shepard

Soil Not Oil: Environmental Justice in an Age of Climate Crisis, by Vandana Shiva

The Soil Will Save Us: How Scientists, Farmers, and Foodies Are Healing the Soil to Save the Planet, by Kristin Ohlson

The Uninhabitable Earth: Life after Warming, by David Wallace-Wells

Water in Plain Sight: Hope for a Thirsty World, by Judith D. Schwartz

Amazing Carbon, a website curated by Christine Jones (amazingcarbon.com)

Recommended Videos

"Cover Crops for Grazing," by Gabe Brown, at the 2016 Grassfed Exchange annual conference

youtube.com/watch?v=tuwwfL2o9d4

"Grazing Down the Carbon: The Scientific Case for Grassland Restoration," by Richard Teague, at Restoring Ecosystems to Reverse Global Warming, a conference hosted by Biodiversity for a Livable Climate, November 22, 2014

youtube.com/watch?v=rhDq_VBhMWg

Green Gold, by John D. Liu

youtube.com/watch?v=YBLZmwlPa8A&list=PL_OCfTZ7 -XBAjq6lXO5ej6_JHRF1iQIM2&index=8&t=0s

How to Green the World's Deserts and Reverse Climate Change, by Allan Savory

youtube.com/watch?v=vpTHi7O66pI&t=0s&list=PL_OCfTZ7 -XBAjq6lXO5ej6_JHRF1iQIM2&index=4

Johnson Su Bioreactor, by David Johnson

youtube.com/watch?time_continue=7&v=DxUGk161Ly8

Living Soil, by the Soil Health Institute

youtube.com/watch?v=ntJouJhLM48

Put Carbon Where It Belongs . . . Back in the Soil, by NOFA/Mass

youtube.com/watch?v=S3rhjqzVrRc&list=PL_OCfTZ7 -XBAjq6lXO5ej6_JHRF1iQIM2&index=3&t=0s

"Regenerative Agriculture: A Solution to Climate Change," by Ben Dobson, at TEDx Hudson (New York), September 27, 2014
youtube.com/watch?v=yp1i8_JFsao

"Regenerating the Land," by Ray Archuleta, at the 2017 Grassfed Exchange annual conference
youtube.com/watch?v=MrBHIuz34k8

"Ronnie Cummins and Dr. Vandana Shiva at the 2015 Regeneration International Conference," by Regeneration International
youtube.com/watch?v=6FslSyYMXvE&list=PL_OCfTZ7-XBAjq6lXO5ej6_JHRF1iQIM2&index=10

Small Scale Farmers Cool the Planet, by Fair World Project
youtube.com/watch?v=GjD8URaGe88&feature=youtu.be

Soil Carbon Cowboys: A Conversation with Peter Byck and Allen Williams, by New America
youtube.com/watch?v=HDuwmnME0T0

The Soil Solution to Climate Change, by Sustainable World Media
youtube.com/watch?v=BxiXJnZraxk

Soil Solutions to Climate Problems, narrated by Michael Pollan, by Center for Food Safety
youtube.com/watch?v=NxqBzrx9yIE&feature=youtu.be

The Soil Story, by Kiss the Ground
youtube.com/watch?v=nvAoZ14cP7Q&feature=youtu.be

Any videos posted on the Facebook page of Biodiversity for a Livable Climate
facebook.com/pg/bio4climate/videos/

Key Organizations (USA)

Regeneration International
Website: regenerationinternational.org

Organic Consumers Association
Website: organicconsumers.org

Sunrise Movement
Website: sunrisemovement.org

Farmers and Ranchers for a Green New Deal
Website: regenerationinternational.org/farmers-ranchers-green-new-deal/

Regenerative Organic Alliance
Website: regenorganic.org

Real Organic Project
Website: realorganicproject.org

Acres USA
Website: acresusa.com

Northeast Organic Farming
Association
Website: nofa.org

American Grassfed Association
Website: americangrassfed.org

Biodiversity for a Livable Climate
Website: bio4climate.org

Regenerate Nebraska
Website: gcresolve.com/regenerate

Regeneration Vermont
Website: regenerationvermont.org

Biodynamic Association
Website: biodynamics.com

Mercola.com
Website: mercola.com

Institute for Agriculture & Trade
Policy
Website: iatp.org

Kiss the Ground
Website: kisstheground.com

New Mexico State University
Institute for Sustainable
Agricultural Research
Website: case.nmsu.edu/case
/pasodelnorteagriculturalworkshops
/documents/Johnson-Su
_Bioreactor_BMP.pdf

Savory
Website: savory.global

Thunder Valley Community
Development Corporation
Website: thundervalley.org

Singing Frogs Farm
Website: singingfrogsfarm.com

Quivira Coalition
Website: quiviracoalition.org

Rodale Institute
Website: rodaleinstitute.org

Patagonia Provisions
Website: patagoniaprovisions.com

Dr. Bronner's
Website: drbronner.com

Baker Creek Heirloom Seeds
Website: rareseeds.com

Southeastern African American
Farmers' Organic Network
(SAAFON)
Website: saafon.org

Savanna Institute
Website: savannainstitute.org

California State University at Chico
Center for Regenerative Agriculture
and Resilient Systems
Website: csuchico.edu
/regenerativeagriculture

National Association of
Conservation Districts
Website: nacdnet.org

Sustainable Food Lab
Website: sustainablefoodlab.org

Fibershed
Website: fibershed.com

Project Drawdown
Website: drawdown.org

The #NoRegrets Initiative
Website: noregretsinitiative.com

California Climate
& Agriculture Network
Website: calclimateag.org

The Land Institute
Website: landinstitute.org

Carbon Farming Solution
Website: carbonfarmingsolution.com

The Carbon Underground
Website: thecarbonunderground.org

Civil Eats
Website: civileats.com

Food and Farm Communications Fund
Website: foodandfarm
communications.org

Main Street Project
Website: mainstreetproject.org

Perennial Farming Initiative
Website: perennialfarming.org

Nature4Climate
Website: nature4climate.org

Breakthrough Strategies & Solutions
Website: breakthroughstrategies
andsolutions.com

Forest Trends
Website: forest-trends.org

Data for Progress
Website: dataforprogress.org

#GreenforAll
Website: greenforall.org

Green Infrastructure Foundation
Website: greeninfrastructure
foundation.org

Green Forests Work
Website: greenforestswork.org

GreenWave
Website: greenwave.org

Honor the Earth
Website: honorearth.org

Carbon Cowboys
Website: carboncowboys.org

Key International and Global Organizations

Regeneration International
Website: regenerationinternational.org

IFOAM-Organics International
Website: ifoam.bio

"4 per 1000" Initiative
Website: 4p1000.org

GRAIN
Website: grain.org

School Strike 4 Climate
(Australia)
Website: schoolstrike4climate.com

Extinction Rebellion
Website: rebellion.earth

International Analog
Forestry Network
Website: analogforestry.org

Millennium Institute
Website: millennium-institute.org

Sustainable Food Trust (UK)
Website: sustainablefoodtrust.org

Netherlands Centre for
Indigenous Peoples
Website: povertyandconservation
.info/en/org/o0165

Healthy Soils Australia
Website: healthysoils.com.au

Environmental Education
Media Project
Website: eempc.org

Fazenda de Toca (Brazil)
Website: fazendadatoca.com.br/

Howard G. Buffett Foundation Centre
for No-Till Agriculture (Ghana)
Website: centrefornotill.org

Carbon Neutral Cities Alliance
Website: usdn.org/public/page
/13/CNCA

Vía Orgánica (Mexico)
Website: viaorganica.org

Regeneration Belize
Website: regenerationinternational
.org/belize
Facebook: facebook.com
/Regeneration-Belize
-703524823104970

Sustainable Harvest International
(Mexico and Central America)
Website: sustainableharvest.org

Shumei International
Website: shumei-international.org

Pasticultores del Desierto (Mexico)
Website: pasticultores
deldesierto.com

Terra Genesis International
Website: terra-genesis.com

International Union for
Conservation of Nature
Website: iucn.org

Global Peatlands Initiative
Website: globalpeatlands.org

Nia Tero
Website: niatero.org

Society for Ecological Restoration
Website: ser.org

Groundswell International
Website: groundswell
international.org

Grassroots Trust (Zambia)
Website: grassrootstrust.com

Amalima (Zimbabwe)
Website: cnfa.org/program
/amalima

Amrita Bhoomi (India)
Website: amritabhoomi.org

Las Cañadas (Mexico)
Website: greenbiz.com
/article/farm-grows-climate
-change-solution

Farmer Field School
Website: fao.org/farmer
-field-schools/en

Better Soils Better Lives (Africa)
Website: bettersoilsbetterlives.org/

Africa Centre for Holistic Management
Website: africacentrefor
holisticmanagement.org

Alliance for Food Sovereignty in Africa
Website: afsafrica.org

Holistic Management International
Website: holisticmanagement.org

La Via Campesina
(International Peasant's Movement)
Website: viacampesina.org/en

Rare
Website: rare.org

Soils for Life (Australia)
Website: soilsforlife.org.au

Vi Agroforestry (Africa)
Website: viagroforestry.org

World Agroforestry
Website: worldagroforestry.org

Great Green Wall Project (Africa)
Website: greatgreenwall.org

Ecosystem Restoration Camps
Website: ecosystem
restorationcamps.org

China Green Foundation
Website: cgf.org.cn/en

African Forest Landscape
Restoration Initiative
Website: AFR100.org

Soil Association (UK)
Website: soilassociation.org

Trees for the Future (Africa)
Website: trees.org

UN Environment Program
Trillion Trees Campaign
PDF about the campaign: un
environment.org/resources/publication
/plant-planet-billion-tree-campaign

International Biochar Initiative
Website: biochar-international.org/

350.org
Website: 350.org/

Regenerative Investing

Confluence Philanthropy
Website: confluence
philanthropy.org

Farmland LP
Website: farmlandlp.com

Indigo: The Terraton Initiative
Website: indigoag.com
/the-terraton-initiative

Iroquois Valley Farmland
Website: iroquoisvalley.com

Propagate Ventures
Website: propagateventures.com

Root Capital
Website: rootcapital.org

RSF Social Finance
Website: rsfsocialfinance.org

Regenerative Agriculture
Investor Network
Website: lifteconomy.com/rain

Shared Capital Cooperative
Website: sharedcapital.coop

NOTES

Introduction

1. For a discussion of why sustainability is an outdated concept, see Ronnie Cummins and André Leu, "From 'Sustainable' to 'Regenerative'—The Future of Food," Organic Consumers Association, October 26, 2015, https://www.organicconsumers .org/essays/%E2%80%98sustainable%E2%80%99-%E2%80%98regenerative %E2%80%99%E2%80%94-future-food.

Chapter 1: Rules for Regenerators

1. For details on how factory farming can be seen as the linchpin of corporate environmental destruction, see Ronnie Cummins and Martha Rosenberg, "Time to Drive Factory Farmed Food off the Market," Organic Consumers Association, September 27, 2016, https://www.organicconsumers.org/essays/time-drive -factory-farmed-food-market.

2. Dr. Joseph Mercola, "Depression Spikes 33 Percent in 5 Years," Mercola.com, May 24, 2018, https://articles.mercola.com/sites/articles/archive/2018/05/24 /depression-spikes-33-percent-in-5-years.aspx.

3. For more information on the "4 per 1000" Initiative, see Ronnie Cummins, "'Four for 1000': A Global Initiative to Reverse Global Warming through Regenerative Agriculture and Land Use," Regeneration International, October 17, 2017, https://regenerationinternational.org/2017/10/17/four-1000-global-initiative -reverse-global-warming-regenerative-agriculture-land-use/.

4. Paul Hawken, ed., *Drawdown: The Most Comprehensive Plan Ever Proposed to Reverse Global Warming* (New York: Penguin Books, 2017), 137–56.

5. Vaselina Petrova, "World Could Win Climate Battle with 100% Renewables Transition by 2050," Renewables Now, January 22, 2019, https://renewablesnow .com/news/world-could-win-climate-battle-with-100-renewables-transition -by-2050-640103/.

6. For a list of investors, corporations, and institutions that have committed to divesting from fossil fuels, see "1000+ Divestment Commitments," Fossil Free Divestment, July 25, 2019, https://gofossilfree.org/divestment/commitments/.

7. Cristina Z. Peppard, "What You Need to Know about Pope Francis's Environmental Encyclical," *Washington Post*, June 16, 2015, https://www.washingtonpost.com /news/acts-of-faith/wp/2015/06/18/what-you-need-to-know-about-pope -franciss-environmental-encyclical/?utm_term=.df391d88279e.

8. Dr. Joseph Mercola, "Regenerative Agriculture—The Next Big Thing," March 27, 2018, https://articles.mercola.com/sites/articles/archive/2018/03/27/regenerative -agriculture-the-next-big-thing.aspx.

9. Thomas Reese, "A Reader's Guide to 'Ladato Si,'" *National Catholic Reporter*, June 25, 2016, https://www.ncronline.org/blogs/faith-and-justice/readers -guide-laudato-si.

10. For more discussion on the role of factory farms in climate change, environmental destruction, and deteriorating public health, see Cummins and Rosenberg, "Time to Drive Factory Farmed Food off the Market."

11. Jack Kittredge, *Soil Carbon Restoration: Can Biology Do the Job?* (Northeast Organic Farming Association/Massachusetts Chapter, August 14, 2015), https:// www.nofamass.org/sites/default/files/2015_White_Paper_web.pdf.

12. Dr. Joseph Mercola, "The Case against Veganism—Carefully Researched Book Spills the Beans," Mercola.com, September 25, 2016, https://articles.mercola .com/sites/articles/archive/2016/09/25/veganism.aspx.

13. For more on the benefits of local organic farming, see Will Allen, Ronnie Cummins, and Kate Duesterberg, "Local and Organic Food and Farming: The Gold Standard," Organic Consumers Association, February 22, 2011, https://www .organicconsumers.org/essays/local-and-organic-food-and-farming-gold-standard.

14. Wendell Berry, *The Art of the Commonplace: The Agrarian Essays*, ed. Norman Wirzba (Berkeley, CA.: Counterpoint, 2002).

Chapter 2: Regeneration

1. James Bruges, *The Biochar Debate* (White River Junction, VT.: Chelsea Green Publishing, 2009), 85–87.

2. David Wasdell, "Climate Dynamics: Facing the Harsh Realities of Now," Apollo-Gaia Project, retrieved August 7, 2019, http://www.apollo-gaia.org /harsh-realities-of-now.html.

3. Wasdell, "Climate Dynamics."

4. Thomas J. Goreau, ed., *Geotherapy: Innovative Methods of Soil Fertility Restoration, Carbon Sequestration, and Reversing CO2 Increase* (New York: CRC Press, 2015), 179, and Eric Toensmeier, *The Carbon Farming Solution* (White River Junction, VT.: Chelsea Green Publishing, 2016), 10.

5. Wayne A. White, *Biosequestration and Ecological Diversity* (New York: CRC Press, 2013), p. 100.

6. Maureen Brown, "Can Regenerative Agriculture Save Our Planet?," *Kiwi Magazine*, July 2017, http://www.kiwimagonline.com/2017/07/can-regenerative -agriculture-save-planet/.

7. Toensmeier, *The Carbon Farming Solution*, 54–55.

8. Luke Smith, "Regenerating Supply at the Growing Edge of the Natural Products Industry," Medium, April 16, 2018, https://medium.com/terra-genesis /regenerating-supply-at-the-growing-edge-of-the-natural-products -industry-e682a17ff0ed.

9. Smith, "Regenerating Supply."

10. Ronnie Cummins, "Regeneration: The Next Stage of Organic Food and Farming— and Civilization," Organic Consumers Association, May 28, 2017, https://www .organicconsumers.org/regeneration-next-stage-organic-food-and-farming -and-civilization.

11. Dr. Joseph Mercola, "The Dorito Effect—The Surprising Truth about Food and Flavor," Mercola.com, June 3, 2018, https://articles.mercola.com/sites/articles /archive/2018/06/03/truth-about-food-and-flavor.aspx.

12. ETC Group, *Who Will Feed Us? The Industrial Food Chain vs. the Peasant Food Web* (October 10, 2017), 12–13, 52, https://www.etcgroup.org/content/who-will -feed-us-industrial-food-chain-vs-peasant-food-web.

13. ETC Group, *Who Will Feed Us?*.

Chapter 3: Grassroots Awareness, Political Mobilization, and Marketplace Demand

1. David Wallace Wells, "The Uninhabitable Earth," *New York Magazine*, July 10, 2017, http://nymag.com/daily/intelligencer/2017/07/climate-change-earth -too-hot-for-humans.html.

2. For more on the DARK Act, see Ronnie Cummins, "Beyond Frankenfoods and the DARK Act: A Grassroots-Powered Revolution," Organic Consumers Association, July 18, 2016, https://www.organicconsumers.org/essays/beyond -frankenfoods-and-dark-act-grassroots-powered-revolution.

3. For a look at the organic supply chain infrastructure and the big-name companies that run it, see the infographic by Dr. Phil Howard in "Who Owns Organic," Cornucopia Institute, retrieved August 7, 2019, https://www.cornucopia.org /who-owns-organic/.

4. Kai Ryssdal, "Processed Foods Make Up 70 Percent of the U.S. Diet," *Marketplace*, March 12, 2013, https://www.marketplace.org/2013/03/12/life/big-book /processed-foods-make-70-percent-us-diet.

5. Tara Parker-Pope, "An Omnivore Defends Real Food," *New York Times*, January 17, 2008, https://well.blogs.nytimes.com/2008/01/17/an-omnivore-defends-real-food/.

6. Elizabeth Royte, "One-Third of Food Is Lost or Wasted: What Can Be Done," *National Geographic*, October 13, 2014, https://news.nationalgeographic.com/news /2014/10/141013-food-waste-national-security-environment-science-ngfood/, and Eillie Anzilotti, "We're Leaving So Much Food on Farms to Just Rot in the

Fields," *Fast Company*, July 24, 2019, https://www.fastcompany.com/90380088
/were-leaving-so-much-food-on-farms-to-just-rot-in-the-fields.

7. "Obesity and Overweight," World Health Organization, February 16, 2018,
   https://www.who.int/news-room/fact-sheets/detail/obesity-and-overweight,
   and "Globally Almost 870 Million Chronically Undernourished—New Hunger
   Report," UN Food and Agriculture Organization, October 9, 2012, http://www
   .fao.org/news/story/en/item/161819/icode/.

8. "100 Years of U.S. Consumer Spending Data for the Nation, New York City,
   and Boston," U.S. Department of Labor, May 2006, https://www.bls.gov/opub
   /uscs/1950.pdf.

9. "How Much Are Households Spending on Food?" Eurostat, European Commission, updated April 12, 2018, https://ec.europa.eu/eurostat/web/products
   -eurostat-news/-/DDN-20181204-1?inheritRedirect=true.

10. "Millennials Dig In: #1 Hobby in America Gets Boost from Interest in Locally
    Grown Foods," Lowe's Companies, April 23, 2014, https://www.prnewswire
    .com/news-releases/gardening-grows-as-millennials-dig-in-1-hobby-in
    -america-gets-boost-from-interest-in-locally-grown-foods-256355551.html.

11. Beth Gardiner, "A Boon for Soil and the Environment," *New York Times*, May 17,
    2016, http://www.nytimes.com/2016/05/18/business/energy-environment/a
    -boon-for-soil-and-for-the-environment.html?ref=energy-environment&_r=0.

12. ETC Group, *Who Will Feed Us?*.

13. ETC Group, *Who Will Feed Us?*.

14. ETC Group, *Who Will Feed Us?*.

15. Katherine Paul, "French Ministry of Agriculture Official, Leading U.S. Soil
    Scientists Outline Plan to Stall Global Warming through Soil Carbon Seques-
    tration," Organic Consumers Association, March 8, 2016, https://www
    .organicconsumers.org/press/french-ministry-agriculture-official-leading-us
    -soil-scientists-outline-plan-stall-global.

Chapter 4: Carbon Farming, Reforestation, and Ecosystem Restoration

1. Toensmeier, *The Carbon Farming Solution*, 10, 54.

2. Bruges, *The Biochar Debate*, 86–87.

3. For information on cover crops and crop rotations, see the resources listed on
   the web page "Cover Crops: Crop Rotations," USDA Sustainable Agriculture
   Research and Education, retrieved August 7, 2019, https://www.sare.org
   /Learning-Center/Topic-Rooms/Cover-Crops/Cover-Crops-Crop-Rotations.

4. For information on the techniques of double digging and keyline plowing, see
   Brian Barth, "Double Digging: How to Build a Better Veggie Bed," *Modern*

Farmer, March 7, 2016, https://modernfarmer.com/2016/03/double-digging/, and Benjamin Falloon, "Keyline Plowing with Compost Tea Application," *Permaculture News*, September 16, 2009, https://permaculturenews.org /2009/09/16/keyline-plowing-with-compost-tea-application/.

5. For information on alley cropping, see "Alley Cropping and Silvopasture," Matter of Trust.org, retrieved August 7, 2019, https://matteroftrust.org/14413/alley -cropping-and-silvopasture.

6. For more details on the production and uses of biochar, see Albert Bates, *The Biochar Solution* (Gabriola Island, B.C.: New Society Publishers, 2010).

7. To understand why native plants are especially good at attracting pollinators, see Kim Eierman, "Native Cultivars vs. Native Plants & Their Attractiveness to Pollinators," Ecobeneficial.com, retrieved August 7, 2019, https://www .ecobeneficial.com/2014/04/native-cultivars-vs-native-plants/.

8. Lara Bryant, "Organic Matter Can Improve Your Soil's Water Holding Capacity," Natural Resources Defense Council, April, 2014, https://www.nrdc.org/experts /lara-bryant/organic-matter-can-improve-your-soils-water-holding-capacity.

9. For more information on permaculture, see "What Is Permaculture?," Permaculture Principles.com, retrieved August 7, 2019, https://permacultureprinciples.com/.

10. For more information on planned rotational grazing, see "What Is Holistic Planned Grazing?," Savory, February 2017, https://www.savory.global/wp -content/uploads/2017/02/about-holistic-planned-grazing.pdf.

11. Dr. Richard Teague, testimony to the U.S. House of Representatives, Committee on Natural Resources, June 24, 2015, https://naturalresources.house.gov/calendar /eventsingle.aspx?EventID=384738.

12. "Facilitating the Regeneration of Grasslands," Savory Global, October 17, 2019, https://www.savory.global/.

13. Nilovna Chatterjee, P. K. Ramachandran, Saptarshi Chakraborty, and Vimala D. Nair, "Changes in Soil Carbon Stocks across the Forest-Agroforest-Agriculture/ Pasture Continuum in Various Agroecological Regions: A Meta-Analysis," *Agriculture, Ecosystems & Environment* 266 (Nov. 1, 2018): 55–67.

14. Boyd Kidwell, "Mob Grazing," Angus Beef Bulletin, March 2010, http://www .angusbeefbulletin.com/ArticlePDF/MobGrazing%2003_10%20ABB.pdf.

15. For more on multispecies grazing, see Lee Rineheart, "Multispecies Grazing: A Primer on Diversity," ATTRA Sustainable Agriculture, October 17, 2019, https://attra.ncat.org/attra-pub/summaries/summary.php?pub=244.

16. Hawken, *Drawdown*, 50–51.

17. Jim Laurie, "Appendix A: Scenario 300," in *Compendium of Scientific and Practical Findings Supporting Eco-Restoration to Address Global Warming*, vol. 2, no. 1 (Biodiversity for a Livable Climate, July 2018), 49.

18. Kim Severson, "At White Oak Pastures, Grass-Fed Beef Is Only the Beginning," *New York Times*, March 11, 2015, https://www.nytimes.com/2015/03/11/dining /at-white-oak-pastures-grass-fed-beef-is-only-the-beginning.html.

19. Toensmeier, *The Carbon Farming Solution*, 31.

20. Laurie, "Appendix A: Scenario 300," 41.

21. Jack Kittredge, *Soil Carbon Restoration: Can Biology Do the Job?* (Northeast Organic Farming Association/Massachusetts Chapter), August 14, 2015, https://www.nofamass.org/sites/default/files/2015_White_Paper_web.pdf.

22. Dr. David Johnson, "The BEAM Approach," presentation given at the Soil Carbon Trading Roadshow 2016, posted by Carbon Link, January 26, 2017, https://www.youtube.com/watch?v=79qpP0m7SaY.

23. Toensmeier, *The Carbon Farming Solution*, 31.

24. Wayne A. White, *Biosequestration and Ecological Diversity* (New York: CRC Press, 2013), 101.

25. White, *Biosequestration and Ecological Diversity*, 101.

26. "The Trillion Tree Declaration," Plant for the Planet, retrieved August 8, 2019, https://www.plant-for-the-planet.org/de/informieren/trillion-tree-declaration.

27. Josh Gabbatiss, "Massive Restoration of World's Forests Would Cancel Out a Decade of CO2 Emissions," *The Independent* (UK), February 16, 2019, https:// www.independent.co.uk/environment/forests-climate-change-co2-greenhouse -gases-trillion-trees-global-warming-a8782071.html.

28. Hawken, *Drawdown*, 110.

29. Jesse Bussard, "Livestock Grazing Can Be a Useful Tool for Fire & Weed Suppres-sion," Gallagher Group, retrieved August 8, 2019, https://am.gallagher.com/us /in-practice/livestock-grazing-can-be-a-useful-tool-for-fire-weed-suppression.

30. Tobias Roberts, "The Best Species for Coppice Forestry," Permaculture Research Institute, September 15, 2017, https://permaculturenews.org/2017/09/15/best -species-coppice-forestry/.

31. Toensmeier, *The Carbon Farming Solution*, 30–32.

32. For more information on alley cropping perennials with annuals, see "Interview with Mark Shepard," Northeast Organic Farming Association, November 2013, https://www.nofamass.org/articles/2013/11/interview-mark-shepard.

33. For more information on biodiverse food forests, see Martin Crawford, "Forest Gardening," Agroforestry Research Trust, retrieved August 8, 2019, https:// www.agroforestry.co.uk/about-agroforestry/forest-gardening/, and Matthew Wilson, "How to Create a Woodland Where Everything Is Edible," *Financial Times*, January 30, 2015, https://www.ft.com/content/db556ddc-a54e-11e4 -ad35-00144feab7de.

34. Toensmeier, *The Carbon Farming Solution*, 112.

35. White, *Biosequestration and Ecological Diversity*, 80.

36. White, *Biosequestration and Ecological Diversity*, 81.

37. Hawken, *Drawdown*, 110.

38. Toensmeier, *The Carbon Farming Solution*, 5.

39. "Wetlands," in *Compendium of Scientific and Practical Findings Supporting Eco-Restoration to Address Global Warming*, vol. 1, no. 1 (Biodiversity for a Livable Climate, July 2017), 36–39, https://bio4climate.org/wp-content/uploads/Compendium-Vol-1-No-1-July-2017-Biodiversity-for-a-Livable-Climate-1.pdf.

40. Laurie, "Appendix A: Scenario 300."

41. John D. Liu, a founding member of Regeneration International, documented large-scale ecosystem restoration projects in China (including the Loess Plateau), Africa (Ethiopia, Rwanda), South America, and the Middle East in his film *Green Gold* (2012), highlighting the enormous benefits for people and planet of "greening deserts." You can find the documentary on YouTube at https://www.youtube.com/watch?v=YBLZmwlPa8A.

42. Issa Sikiti da Silva, "Great Green Wall Brings Hope, Greener Pastures to Africa's Sahel," Interpress Service, June 11, 2018, http://www.ipsnews.net/2018/06/great-green-wall-brings-hope-greener-pastures-africas-sahel.

Chapter 5: Politics and Public Policy

1. Cody Nelson, "What's the 'Green New Deal' and Why Do Environmentalists Want It?," Minnesota Public Radio, November 19, 2018, https://www.mprnews.org/story/2018/11/19/green-new-deal-omar-ocasio-cortez.

2. Tim Dickinson, "Getting to the Bottom of the Green New Deal," *Rolling Stone*, January 7, 2019, https://www.rollingstone.com/politics/politics-news/green-new-deal-explained-775827/.

3. Dickinson, "Getting to the Bottom of the Green New Deal."

4. Jeremy Bloom, "Here's the Full Text of Congress' Green New Deal Resolution, Introduced by Rep. Alexandria Ocasio Cortez," Clean Technica, February 8, 2019, https://cleantechnica.com/2019/02/08/heres-the-full-text-of-congress-green-new-deal-resolution-introduced-by-rep-alexandra-ocasio-cortez/.

5. For more info on a proposed Green New Deal in the UK, see Matthew Taylor, "Labour Would 'Radically Transform Economy' to Focus on Climate Change," *The Guardian*, December 24, 2018, https://www.theguardian.com/politics/2018/dec/24/labour-government-tackle-climate-change.

6. Naomi Klein, "The Game-Changing Promise of a Green New Deal," *The Intercept*, November 27, 2018, https://theintercept.com/2018/11/27/green-new-deal-congress-climate-change/.

7. Steve Hanley, "Ho Hum. 6th IPCC Report Says World Is Headed for Existential Crisis. What Else Is New?," Clean Technica, September 27, 2018, https://cleantechnica.com/2018/09/27/ho-hum-6th-ipcc-report-says-world-is-headed-for-existential-crisis-what-else-is-new/.

8. For more on best practices in sequestering soil carbon, see André Leu, "Reversing Climate Change through Regenerative Agriculture," Regeneration International, October 9, 2018, https://regenerationinternational.org/2018/10/09/reversing-climate-change-through-regenerative-agriculture/, and Han de Groot, "The Best Technology for Fighting Climate Change Isn't a Technology," *Scientific American*, December 5, 2018, https://blogs.scientificamerican.com/observations/the-best-technology-for-fighting-climate-change-isnt-a-technology/.

9. For a good overview on soil carbon sequestration, see Anne-Marie Codur, Seth Itzkan, William Moomaw, Karl Thidemann, and Jonathan Harris, "Hope below Our Feet: Soil as a Climate Solution," Global Development and Environment Institute, Tufts University, April 2017, http://www.ase.tufts.edu/gdae/Pubs/climate/ClimatePolicyBrief4.pdf.

10. For more info on GHG emissions and air pollution from factory farms, see "Fact Sheet: Air Pollution from Factory Farms," Environmental Integrity, retrieved August 8, 2019, http://environmentalintegrity.org/wp-content/uploads/FACT-SHEET-on-air-pollution-from-factory-farms.pdf, and Ronnie Cummins and Martha Rosenberg, "Monsanto's Evil Twin: Disturbing Facts about the Fertilizer Industry," Organic Consumers Association, April 5, 2016, https://www.organicconsumers.org/essays/monsanto%E2%80%99s-evil-twin-disturbing-facts-about-fertilizer-industry.

11. Lisa Held, "Cory Booker Wants to Pay Many More Farmers to Practice Carbon Farming," Civil Eats, last modified August 8, 2019, https://civileats.com/2019/08/08/cory-booker-wants-to-pay-many-more-farmers-to-practice-carbon-farming/.

12. Elizabeth Henderson, "Why Sustainable Agriculture Should Support a Green New Deal," Organic Consumers Association, January 1, 2019, https://www.organicconsumers.org/news/why-sustainable-agriculture-should-support-green-new-deal.

13. Henderson, "Why Sustainable Agriculture Should Support a Green New Deal."

14. Jesse Colombo, "Here's Why More American Farms Are Going Bankrupt," *Forbes*, November 29, 2018, https://www.forbes.com/sites/jessecolombo/2018/11/29/heres-why-more-american-farms-are-going-bankrupt/#14afcf6865a7, and Nick Meyer, "10 Favorite Organic Food Makers That Are Now Owned by Huge Corporations," Alt Health Works, October 8, 2018, https://althealthworks.com/6531/10-favorite-organic-food-makers-that-are-now-owned-by-huge-corporations/.

15. Devon Downeysmith, "On Clean Energy, Portland Takes the Initiative," Climate Solutions, November 8, 2018, https://www.climatesolutions.org/article /1541788536-clean-energy-portland-takes-initiative.

16. Emily Atkin, "Pentagon: Global Warming Poses 'Immediate Risk' to National Security," Think Progress, October 14, 2014, https://www.thinkprogress.org /climate/2014/10/14/3579338/pentagon-global-warming-national-security/.

17. Bill Becker, "National Security Begins at Home," *Solutions* (online), January 2012, https://www.thesolutionsjournal.com/article/national-security-begins-at-home/.

18. Fen Montaigne, "Amory Lovins Lays Out His Clean Energy Plan," Yale Environment 360, February 20, 2012, https://e360.yale.edu/features/amory_lovins _clean_energy_guru_presents_his_master_plan.

19. "Four for 1000: Soils for Food Security and Climate," Regeneration International, retrieved August 8, 2019, http://www.regenerationinternational.org/4p1000.

20. Dr. Richard Teague, at the Regeneration International press conference at the National Press Club in Washington, D.C., March 8, 2016. Quote from the author's audio recording at the press conference.

21. Andy Coghlan, "Warming Arctic Could Be behind Heatwave Sweeping Northern Hemisphere," *New Scientist*, July 24, 2018, https://www.newscientist .com/article/2174889-warming-arctic-could-be-behind-heatwave-sweeping -northern-hemisphere/.

22. Yeva Nersisyan and L. Randall Wray, "How to Pay for the Green New Deal," Levy Institute of Bard College, May 2019, http://www.levyinstitute.org/pubs/wp 931.pdf.

23. "The Cost of US Wars Then and Now," Norwich University Online, retrieved August 8, 2019, https://online.norwich.edu/academic-programs/masters /military-history/resources/infographics/the-cost-of-us-wars-then-and-now.

24. Brian Wang, "World GDP Forecasts for 2030," Next Big Future, January 14, 2019, https://www.nextbigfuture.com/2019/01/world-gdp-forecasts-for-2030.html.

25. Brittany de Lea, "How Much AOC's Green New Deal Could Cost the Average American Household," Fox News Network, July 30, 2019, https://www .foxbusiness.com/economy/aoc-green-new-deal-cost-american-household.

26. Jessica McDonald, "How Much Will the 'Green New Deal' Cost?," FactCheck .org, March 14, 2019, https://www.factcheck.org/2019/03/how-much-will-the -green-new-deal-cost/.

27. Erik Sherman, "U.S. Health Care Costs Skyrocketed to $3.65 Trillion in 2018," *Fortune*, February 21, 2019, https://fortune.com/2019/02/21/us-health-care-costs-2/.

28. Rabah Kamal, Bradley Sawyer, and Daniel McDermott, "How Much Is Health Spending Expected to Grow?," Kaiser Family Foundation, March 12, 2019, https:// www.healthsystemtracker.org/chart-collection/much-health-spending-expected -grow/#item-health-spending-projections-now-lower-previous-projections_2016.

29. Zach Faria, "Sanders: Medicare for All Will Cost $40 Trillion Over 10 Years," *Washington Free Beacon*, July 16, 2019, https://freebeacon.com/politics/sanders -medicare-for-all-will-cost-40-trillion-over-10-years/.

30. Lisa Friedman, "Bernie Sanders's 'Green New Deal': A $16 Trillion Climate Plan," *New York Times*, updated August 28, 2019, https://www.nytimes.com /2019/08/22/climate/bernie-sanders-climate-change.html.

31. Ellen Brown, "The Secret to Funding a Green New Deal," Truthdig, March 19, 2019, https://www.truthdig.com/articles/the-secret-to-funding-a-green -new-deal/.

32. Brown, "The Secret to Funding a Green New Deal."

33. Zach Carter, "Stephanie Kelton Has the Biggest Idea in Washington," *Huffington Post*, May 21, 2018, https://www.huffpost.com/entry/stephanie-kelton-economy -washington_n_5afee5eae4b0463cdba15121.

34. Stephanie Kelton, Andres Bernal, and Greg Carlock, "We Can Pay for a Green New Deal," *Huffington Post*, November 30, 2018, https://www.huffpost.com /entry/opinion-green-new-deal-cost_n_5c0042b2e4b027f1097bda5b.

35. For more information on Regeneration and the false solution of geo-engineering, see Vandana Shiva and Kartikey Shiva, *The Future of Our Daily Bread: Regeneration or Collapse?* (New Delhi: Navdanya International, 2018), https:// seedfreedom.info/wp-content/uploads/2018/11/The-Future-of-Our-Daily -Bread-_-LowRes-_-19-11-2018-REVISED.pdf, and "We Need Regenerative Farming, Not Geoengineering," Community Solutions, May 11, 2016, https:// www.communitysolution.org/blog/2016/5/11/we-need-regenerative-farming -not-geoengineering?mc_cid=4ae65c18e9&mc_eid=4400a49ac6.

Chapter 6: Public and Private Investment

1. Michael Shuman, *Local Dollars, Local Sense* (White River Junction, VT.: Chelsea Green Publishing, 2012), xii–xiii.

2. Shuman, *Local Dollars,* 4, xx.

3. Shuman, *Local Dollars,* xx.

4. ETC Group, *Who Will Feed Us?*.

5. "Soil Wealth Investment in Regenerative Agriculture across Asset Classes: New Report Lays Out Path Forward for Investment in Regenerative Agriculture," July 17, 2019, https://rfsi-forum.com/barriers-and-opportunities-to-regenerative -ag-investment-uncovered-in-new-report/.

6. "Global Fossil Fuel Divestment Movement Reaches $6.24 Trillion in Assets under Management, 120x Increase from Four Years Ago, Report Says," 350.org, September 11, 2018, https://www.commondreams.org/newswire/2018/09/11 /global-fossil-fuel-divestment-movement-reaches-624-trillion-assets-under.

7. ETC Group, *Who Will Feed Us?*

8. Shuman, *Local Dollars*, 3.

9. "What Is Regenerative Agriculture?," The Carbon Underground, February 16, 2017, https://2igmzc48tf4q88z3o24qjfl8-wpengine.netdna-ssl.com/wp-content /uploads/2017/02/Regen-Ag-Definition-2.23.17-1.pdf.

10. For more information about climate-smart agriculture and zero budget natural farming, see Peter Newell, Jennifer Clapp, and Zoe W. Brent, "Will 'Climate Smart Agriculture' Serve the Public Interest—or the Drive for Growing Profits for Private Corporations?," *The Ecologist*, January 19, 2018, https://theecologist .org/2018/jan/19/will-climate-smart-agriculture-serve-public-interest-or -drive-growing-profits-private, and Shiva and Shiva, *The Future of Our Daily Bread*, 47–53.

11. For information on Big Data as a false solution to climate change and improving farmer livelihoods, see Tim McDonnall, "Monsanto Is Using Big Data to Take Over the World," *Mother Jones*, November 19, 2014, https://www.motherjones .com/environment/2014/11/monsanto-big-data-gmo-climate-change/.

12. "Back to Grass: The Market Potential for U.S. Grassfed Beef," Stone Barns Center for Food and Agriculture, April 2017, https://www.stonebarnscenter.org /wp-content/uploads/2017/10/Grassfed_Full_v2.pdf.

13. For information on private sector investing in regenerative agriculture, see Devon Thorpe, "How Investing in Regenerative Agriculture Can Help Stem Climate Change Profitably," *Forbes*, December 12, 2018, https://www.forbes.com/sites /devinthorpe/2018/12/12/how-investing-in-regenerative-agriculture-can-help -stem-climate-change-profitably/#30fa95373e5c.

14. "National Beer Sales & Production Data," Brewers Association for Small and Craft Brewers, retrieved August 9, 2019, https://www.brewersassociation.org /statistics/national-beer-sales-production-data/.

15. Steve Edgerton, "10 Craft Breweries Using Millets and Sorghum," Food Tank, October 2018, https://foodtank.com/news/2018/10/10-craft-breweries -using-millets-and-sorghum/.

16. "U.S. Legal Cannabis Market Projected to Triple by 2025 to $25 Billion," New Cannabis Ventures, April 20, 2018, https://www.newcannabisventures.com /u-s-legal-cannabis-market-projected-to-triple-by-2025-to-25-billion/.

Chapter 7: The Global Road to Regeneration

1. "Summary for Policymakers of IPCC Special Report on Global Warming of 1.5°C Approved by Governments," Intergovernmental Panel on Climate Change, October 8, 2018, https://www.ipcc.ch/2018/10/08/summary-for-policymakers -of-ipcc-special-report-on-global-warming-of-1-5c-approved-by-governments/.

Chapter 8: Roadmap to Regeneration in the United States, 2020–2030

1. "We Are Sunrise," Sunrise Movement, retrieved August 9, 2019, https://www. sunrisemovement.org/, and "Join the Rebellion," Extinction Rebellion, retrieved August 9, 2019, https://rebellion.earth/.

2. Olivia Rosane, "81% of Voters Support a Green New Deal, Survey Finds," EcoWatch, December 18, 2018, https://www.ecowatch.com/green-new-deal -voter-support-2623737355.html.

3. Zoya Teirstein, "Poll: The Green New Deal Is as Popular as Legalizing Weed," *Grist*, July 22, 2019, https://grist.org/article/poll-the-green-new-deal-is-as -popular-as-legalizing-weed/.

4. Laurie, "Appendix A: Scenario 300," 49.

5. "Inventory of U.S. Greenhouse Gas Emissions and Sinks," U.S. Environmental Protection Agency, retrieved August 9, 2019, https://www.epa.gov/ghgemissions /inventory-us-greenhouse-gas-emissions-and-sinks.

6. Damien Carrington, "'Brutal News': Global Carbon Emissions Jump to All-Time High in 2018," *The Guardian*, December 5, 2018, https://www.theguardian.com /environment/2018/dec/05/brutal-news-global-carbon-emissions-jump-to-all -time-high-in-2018.

7. Duncan Clark, "Which Nations Are Most Responsible for Climate Change?," *The Guardian*, April 21, 2011, https://www.theguardian.com/environment/2011 /apr/21/countries-responsible-climate-change.

8. Johan Falk, Owen Gaffney, A. K. Bhowmik, et al., *Exponential Roadmap*, version 1.5 (Sweden: Future Earth, September 2019), https://exponentialroadmap.org /wp-content/uploads/2019/09/ExponentialRoadmap_1.5_20190919 _Single-Pages.pdf.

9. "Sources of Greenhouse Gas Emissions," U.S. Environmental Protection Agency, retrieved August 9, 2019, https://www.epa.gov/ghgemissions/sources -greenhouse-gas-emissions.

10. Ronnie Cummins, "The 9-Percent Lie: Why Are the USDA and EPA Hiding the Fact That Half of All U.S. Greenhouse Gas Emissions Come from Industrial Food, Farming, and Land Use?," Organic Consumers Association, July 17, 2019, https://www.organicconsumers.org/nine-percent-lie.

11. "Which Countries Have a Net Zero Carbon Goal?," Climate Home News, June 14, 2019, https://www.climatechangenews.com/2019/06/14/countries-net -zero-climate-goal/.

12. Mark Z. Jacobson, Mark A. Delucchi, Guillaume Bazouin, et al., "100% Clean and Renewable Wind, Water, and Sunlight (WWS) All-Sector Energy

Roadmaps for the 50 United States," *Energy and Environmental Science* 7 (2015): 2093–117, https://pubs.rsc.org/en/content/getauthorversionpdf/C5EE01283J.

13. Zachary Shahan, "Low Costs of Solar Power & Wind Power Crush Coal, Crush Nuclear, & Beat Natural Gas," Clean Technica, December 25, 2016, https://cleantechnica.com/2016/12/25/cost-of-solar-power-vs-cost-of-wind-power-coal-nuclear-natural-gas/, and Oliver Milman, "'Coal Is on the Way Out': Study Finds Fossil Fuel Now Pricier Than Solar or Wind," *The Guardian*, March 25, 2019, https://www.theguardian.com/environment/2019/mar/25/coal-more-expensive-wind-solar-us-energy-study.

14. Jamie Condliffe, "By 2040, More Than Half of All New Cars Could Be Electric," MIT Technology Review, July 6, 2017, https://www.technologyreview.com/s/608231/by-2040-more-than-half-of-all-new-cars-could-be-electric/.

15. Falk et al., *Exponential Roadmap*, 7, 52–53.

16. Falk et al., *Exponential Roadmap*.

17. Falk et al., *Exponential Roadmap*, and Per-Anders Enkvist and Per Klevnäs, *The Circular Economy: A Powerful Force for Climate Mitigation* (Sweden: Material Economics, June 2018), https://media.sitra.fi/2018/06/12132041/the-circular-economy-a-powerful-force-for-climate-mitigation.pdf.

18. "Inventory of U.S. Greenhouse Gas Emissions and Sinks," U.S. Environmental Protection Agency, retrieved August 9, 2019, https://www.epa.gov/ghgemissions/inventory-us-greenhouse-gas-emissions-and-sinks.

19. Dave Merrill and Lauren Leatherby, "Here's How America Uses Its Land," Bloomberg LP, July 31, 2018, https://www.bloomberg.com/graphics/2018-us-land-use/.

20. Emily Payne, "Dr. Richard Teague: Regenerative Organic Practices 'Clean Up the Act of Agriculture,'" AgFunder News, June 21, 2019, https://agfundernews.com/dr-richard-teague-regenerative-organic-practices-clean-up-the-act-of-agriculture.html.

21. Jack Kittredge, "Soil Carbon Restoration: Can Biology Do the Job?," Northeast Organic Farming Association, August 14, 2015, 8–9, https://www.nofamass.org/sites/default/files/2015_White_Paper_web.pdf.

22. John J. Berger, "Can Soil Microbes Slow Climate Change?," *Scientific American*, March 26, 2019, https://www.scientificamerican.com/article/can-soil-microbes-slow-climate-change/.

23. Sonja N. Oswalt, W. Brad Smith, Patrick D. Miles, and Scott A. Pugh, "Forest Resources of the United States, 2012: A Technical Document Supporting the Forest Service Update of the 2010 RPA Assessment," U.S. Department of Agriculture, retrieved August 9, 2019, https://www.srs.fs.usda.gov/pubs/gtr/gtr_wo091.pdf.

24. Joseph E. Fargione, Steven Bassett, Timothy Boucher, et al., "Natural Climate Solutions for the United States," *Science Advances* 4, no. 11 (November 2018), https://advances.sciencemag.org/content/4/11/eaat1869.

25. Jean-Francois Bastin, Yelena Finegold, Claude Garcia, et al., "The Global Tree Restoration Potential," *Science* 365, no. 6448 (July 2019), https://science .sciencemag.org/content/365/6448/76.full.

26. Jad Daley, "Let's Reforest America to Act on Climate," American Forests, February 26, 2019, https://medium.com/@AmericanForests/lets-reforest-america-to -act-on-climate-1c46ae54acb1.

27. Laurie, "Appendix A: Scenario 300," 53–56.

INDEX

ABOUT THE AUTHOR

R onnie Cummins is founder and director of the Organic
Consumers Association (OCA), a nonprofit, US-based
network of over two million consumers, dedicated to
safeguarding organic standards and promoting a healthy, just,
and regenerative system of food, farming, and commerce. The
OCA's primary strategy is to work on national and global
campaigns promoting health, justice, and regeneration that
integrate public education, marketplace pressure, media work,
litigation, and grassroots lobbying. Cummins is co-editor of
OCA's website and newsletter, *Organic Bytes* (260,000 subscrib-
ers), as well as content coordinator of OCA's two mass social
media sites, OCA and Millions Against Monsanto. Cummins
also serves on the steering committee of Regeneration International and OCA's Mexico affiliate, Vía Orgánica.